A Convenient Lie

Common Sense Talk About Climate Change

Quotable Quotes by DJ Dana:

"The debate will be over when common sense wins over the day and liberal fear mongering evaporates into the night."

"Over reacting to global temperature changes are an emotionally based event, not a logical one."

"Liberals are masterful at avoiding the consequences by hiding behind their good intentions."

"Even if you double the size of a mite, you still can't see it."

"Past performance does sometimes indicate future results."

"Knowledge is power but common sense can save you a lot of money."

"Common sense rules the day in my life, how about yours?"

A Convenient Lie

Common Sense Talk About Climate Change

...

DJ Dana, Author and Thinker

EAN-13: 978-1-45287-156-1

ISBN-10: 1-4528-7156-6

Contents

Preface

"A Convenient Lie" is obviously a retort to Al Gore's infamous book, "An Inconvenient Truth," and is the answer to the whole global climate change groundswell now invading the public consciousness. This is a counter reaction to a new group of global alarmists seeking to scare you out of your SUV and your hard earned tax dollars. Moreover, it is a counter strike to the ongoing ideological battle for the hearts and minds of all breathing human beings.

I wanted to give a very different and optimistic outlook to the unsuspecting people who are being propagandized and inundated with doom and gloom scenarios in regard to this evolving phenomenon called *global climate change*. No doubt most Americans are going about their daily lives in the same manner, attempting to eke out a living in this all-consuming, highly regulated, restrictive, taxing time we live in today.

Consider yourself warned! There is a small, but powerful group of liberals wanting to deprive you of your sovereignty in the name of planet Earth. Their mantra sounds like this, "we (the liberals) need to control everyone's emissions in order to save the planet from ourselves (rather you) and the best solution is for us (the liberals) to tax and regulate everyone who pollutes (or breathes), until our mission (extortion) of saving the planet is completed (day after never)."

First thing you need to ask yourself is, "when has the government ever completed a successful social mission?" Never, they must always grow because the problems get worse and more is needed (taken) to continue the mission. You need to beware the consequences of government action before signing onto the next *New Deal* or what I like to call a *Raw Deal*.

Al Gore will implore you to join his self-proclaimed crusade. If you fall for this act, you will be setting the stage for decades of *in your face* environmental activism and more pompous government officials, less personal choices and no positive results. As I will expose throughout this book, environmentalists and the government have already perpetrated similar scams on us with irreconcilable results.

Be concerned about this radical group of "Greenies" who are attempting to deprive you of your autonomy, but don't be scared by their claims. My personal mission is to reassure sane people that the sky is not falling, nor burning up. I most likely will not win the Nobel Prize, but that is not my goal. However, I will sleep easier at night knowing that I comforted my fellow humans, allowing them to see through the smoke and mirrors befalling us by overzealous liberal activists.

Next, it is important to understand the decrier's motivations, especially when they are seeking to increase taxes, redistribute wealth and alter your standard of living. I lay out the ulterior motives behind the calls to stop our way of life and begin rationing our resources. My motivation is simple: I detest fear-mongering charlatans with half-baked studies, no viable solutions, armed only with predictions of doom and gloom, designed to take away our very freedom!

Furthermore, a more powerful central government is not the answer. Using common sense and real life lessons, I will entice you to agree that free markets and individual enterprise are better equipped to solve problems. Governments are not as altruistic as they like to portray themselves.

Finally, Dr. Covey taught me that every situation, every event has multiple perspectives. That is a human ability - the thinking man's asset; or the complacent person's liability, depending how one hones the mind. Because one man's junk is another man's treasure, it's time a different perspective is presented. The debate is not over!

Acknowledgements

These are my original thoughts and reasoning's. I drew inspiration and insight from talk show sensations Rush Limbaugh and Glenn Beck. Obviously, my passion for truth could not have been so driven if not for the parody of Al Gore Jr. and his documentary. The sarcasm (nearly as witty as Ann Coulter's) is my way of exposing the disingenuous claims being promoted by the left. The puns, often subtle, are like hidden treasures for the careful reader to find.

I owe a special debt of thanks to my loving wife who supports me through thick and thin. My college bound son Brian also collaborated on the cover. In addition, I took advice and edits from my best friend George Johnson, sister-in-law Barbara Dana and many other friends and family... Thanks to all for your support.

Chapter 1   Algor the Caveman

The story of Algor the caveman was derived from my vast research on prehistoric and historic man and how it relates to today's modern environmental movement. Although I only received a history minor from a state college, I am confident that the story presented here is as accurate and true to the times of early man as any of today's mainstream reporting.

The Era of Algor is shortly after the milestone discovery and harness of early mankind's most significant resource – Fire. With the quest for fire over, how would people utilize this powerful tool to further mankind? Early struggles are unearthed here, the rest is history.

Little do people know about this era, as my research will enlighten you to a very tumultuous and defining time in our history! Most people believe that after fire was discovered, man was on easy street. This was far from the reality, as this new powerful resource, which brought heat, cooking, and a weapon for defense, also brought fear. Yes, fear of change, fear of progress, like today, lurks in the minds of some who want to return to a simpler yet laborsome time.

My research uncovered a pivotal point in man's evolution. Shortly after the advent of fire, we entered a period I dubbed – *man's first Dark Age.* As the population declined, man's very own existence was threatened until only the strongest and smartest remained.

Just after the Holocene Ice Age, as the ice began its natural retreat, a leader named Algor and his united tribes worried incessantly about the future and what their place in it would be.

At this point, man had enough crude tools and weapons to acquire fur to keep them warm. In fact, the fur industry was the most powerful influence on the political scene at this time.

From studying the abundance of tree rings, I have learned that this was a bountiful time for plants and thus, animals too. With an abundance of wood, the use of fire spread to every cave in the region. From cave drawings, it was plain to see that all of Algor's tribes used fire to heat their caves and cook their meat. There is even circumstantial evidence that certain advanced tribes used fire to produce medicines and other inventions like torches to keep predatory animals away.

So now, the re-creation of the story of Algor and his united tribes begins with a pow wow of the leaders discussing their concerns for this fledgling community of men and their perceived biggest threat - the unpredictable and mysterious environment.

The Story of Algor

As Algor nervously awaits the leader of the united tribes to discuss with him his newly appointed role as "Algor, Veep of the United Tribes," he ponders, "What can me do to help this great community of man?" It was a calamitous year, as the snow and ice, which everyone was used to, seemed to be dwindling. What to do about it was foremost on the tribal leader's minds. The tribes sometimes needed reassurance from their leaders to keep them from worrying and allow them to go about their business of survival and procreation.

Algor was excited because, until now, he had never gotten to participate in any real policy decisions. He desperately wanted to prove that he could lead the tribes. Secretly, Algor wanted to be "the leader" for as long as he could remember. Even though he had a big cave and nice furs, something was missing in his life.

Just then, Algor looks up to see the tall and virile leader make his grand entrance.

"Me Bilk, me leader of all cavemen and cavewomen." Algor stands and replies, "Me Algor, me Veep thanks to you."

The two hug and sit down by the fire to talk. Bilk begins by affirming what Algor was thinking.

"Me concerned all the snow and ice melting." Biting his lower lip, Bilk states, "Me concerned the tribes scared. Me need solution." Algor nodding his head, feeling Bilk's pain asks, "What me can do?" With puppy dog eyes, Bilk looks at Algor and says, "Me need you to go to all tribes of the land and discover problem." Algor ponders to himself, "If I ever want to be top dog someday, I must discover ice problem." Algor, concerned but invigorated, replies, "Me will do it for you and the tribes."

Algor's Journey begins

The very next day Algor sets out in earnest with the map of the kingdom Bilk has given him, marked with the caves of all the tribal wise-men. Unlike prior days, it is a warm, sunshiny day, and Algor is excited to prove himself a worthy leader by solving the mystery of the melting ice.

It isn't too long before Algor comes to the first cave on his map. Anxiously, Algor knocks on the cave door. A grumpy old man with dirty hands answers…

"Hello, me Algor Veep of all tribes," he stutters.
"Me Metro, man of this cave," the crotchety man grumbles.
Algor trying to soothe the old man's temperament says, "Me told you are important man in this tribe."
A stunned Metro clamors, "Me in charge of trees."
A very impressed Algor thinks to himself that this must be a respected member of the tribe.
Metro continues, "Me in charge of gathering all the firewood. Me hate job!"
Getting right to the point, Algor states, "Me on mission to discover why snow and ice melting."
Metro now becomes impatient, "Me have no idea why. Me gotta go. The kids play in fire pit again and get ash all over cave. Wife tell me to clean up or else!"
Hearing a screeching female voice yell, "Metro, stop gabbing and get back in here,"
Algor decides to move on right after making a mental note about Metro's *ash* problem.

Still energized, Algor is on his way to the next cave on the map. With the overhead, Algor stops to get a drink from a nearby stream. "Man, Sun is hot!" Algor thinks to himself that he should have worn his lighter fur, but it is too soon to return home. Everyone, including Bilk, is counting on him. "Must discover problem," Algor mumbles.

The Breakthrough

Soon Algor arrives at the next cave representing another tribe in the community. Above the entrance, a colorful painting is staring back at him. It is beautiful yet mysterious. Algor is not sure what to make of it having never seen a color picture before. With hopes of gaining more insight, he knocks on the cave door.

With eagerness, Algor greets the caveman. "Me Algor. Me on a mission for our grand leader."
The man stands there silent for a moment, as Algor notices he has ink stained hands.
"Me Painter. Me read old cave drawings and make new ones".
Algor impressed, boldly asks, "What do old drawings say about ice?"
Painter smiles and says, "Come see yourself."

Algor and the Painter venture further into the dark cave. Unfazed, Algor is used to being in the dark.
Painter hands Algor another torch and lights it for him. With the light, Algor can now see the several sets of distinct cave drawings. As Painter shows Algor the old drawings, he notices that there isn't a symbol for fire.

Algor inquires of Painter, "Why no symbol for fire in old drawings?"
Painter answers, "This before fire invented."
Algor follows up with, "What about ice before fire?"
Painter confidently answers, "Lots of ice and snow before fire." as he points to the proof on the dank cave wall.

Algor, still puzzled, knows this is a breakthrough but is not sure why just yet. Algor thanks Painter for his insight and exits toward the next tribe. With plenty of daylight left, Algor meanders down the path mulling over all that he has learned thus far. In no time at all, Algor reaches the furthest tribe on the map. He hopes for the wisest of men to help him solve his riddle.

Tip of the Iceberg

Algor stumbles upon the most magnificent cave he has ever seen. It is twice the size of his cave. "Someone of extreme importance must live here," Algor thinks to himself. Before Algor can get to the door, a man with a white beard and long gray hair comes running out of the cave screaming in pain.

A startled Algor, yells, "What happen! What can me do to help?"
The man shouts, "Me burnt finger in fire, and it really hurts!"
Algor, not sure what to do, instinctively reaches down and grabs a chunk of ice. "Here, put ice on your finger," Algor orders, hoping that will calm him down.
The man grabs the ice and applies it to his burnt finger. Suddenly, a sigh of relief fills the man's face. Algor feels good that he was able to help.
The man now calmed down says, "Thanks. Who are you anyway?"
"Me Algor. Me want to know why ice melting," replies Algor.
With ice still on his finger and not even hearing Algor's question, the man blurts out, "Fire bad, ice good!"
Algor now begins to think of how he can prevent this from happening to other people?
Algor smiles and concludes, "We need warning symbol for fire."
The man nods in agreement and sits down to nurse his wound.

Algor is happy now that he definitely has something to report back to Bilk. He is more encouraged than ever that he will succeed at his mission.

The Consensus Grows

More eager than ever, Algor studies the map to find the shortest route to his destination. Over a hill and through some thick brush, the trek seems to be getting more difficult. Algor arrives beside a large riverbed, but fascinatingly the water is only a small stream where a raging river used to be. Undaunted, Algor knows he must proceed or he will have no chance to be leader. Eventually, Algor arrives at the next cave, a small one, but with lots of colorful plants and flying insects buzzing around.

With enthusiasm, Algor greets the caveman, "Me Algor. Me on a mission to be leader, I mean, for our leader Bilk."
Cautiously, the tall skinny caveman replies, "Me Neander. Me study moths."
"Moths?" asks Algor.
"Yes, moths. The ugly butterflies that are attracted to light," Neander somberly claims.
Algor thinks for a moment, "Oh yes, me see them at night around the cave fire. Why do you seem sad?" Algor wonders aloud.
"All the moths burned up in fire!" cried Neander holding back his tears.

Still stunned by the report, Neander tells Algor, "The bat population is down 10% because no moths to eat."
With Algor's ire up, he exclaims, "Down 30%?" "I will certainly pass on this vital information."
Not wanting to see another caveman cry, Algor gathers his thoughts and heads up the river.

Along this stretch of green land, Algor enjoys the splendors of nature in all its glory. Algor looks up and sees a flock of birds and wonders why the birds always seem to be flying south.

"I wonder what is south of here that makes them all want to leave this place?"
Algor ponders the possibilities.

The Final Leg

Algor, now tired from his long journey, looks anxiously for the final cave on his map. He wants to press on as he feels close to solving the dilemma with which he was tasked.

"Everyone will be so proud of me," he thinks to himself. But wait, "What if no one believes me?" he begins to doubt himself. Perplexed by the mere thought, Algor consoles himself that he must have incontrovertible proof that he is right.
"Me can do it!" Algor convinces himself.
With a newfound confidence, Algor hunts down the final cave on his map. Algor hopes this will be the final piece of the puzzle so he can go home in peace and claim his prize.

Before Algor can knock, a deep voice announces, "Me hungry." Instinctively, Algor steps back and explains to the burly caveman, "Sorry, I have no more food. I am on a mission from our powerful leader"."
"Me Keoto. Me hungry!" he states louder.
Before Algor could ask another question, the big caveman opens his mouth. "Got a sandwich?"
 "Me not invented that, but I am working on a way to network united tribes without leaving my cave," replies Algor.
Keoto puzzled at that invention, just rubs his belly.
Algor growing frustrated, inquires, "Why you so hungry?"
Keoto replies, "Wife burned rabbit stew again. Fire too hot!"
Algor, getting nowhere with Keoto, politely excuses himself.
Algor leaves quickly before Keoto sniffs out the sandwich he has hidden in his fur coat.

Now tired from his long journey, Algor notices the Sun is setting. So close to the answer, but with no final resolution, he is disappointed. Continuing his journey, he reaches a large forest and begins to get cold in the shade. With the Sun setting, he finds a soft patch of moss to bed down for the night; polishing off his last bit of sandwich before falling asleep in the dark damp woods.

Algor begins to dream about Bilk no longer bossing him around and a crowd of people cheering him as he walks out of his cave. Suddenly, Algor wakes up and realizes he was only dreaming. He rolls over on his back and looks up at the thousands of stars in the night sky. He remembers all that the tribesmen have taught him on this very adventurous day.

The Mission Is Over

In the morning, the warm sunshine beats down on Algor, waking him up to the new day. If he goes home now, what will he say? Remembering his first encounter, the timid Metro man had an ash problem, and everyone knows fire causes ash. Proud he did not show Painter his fear of the dark, Algor vividly sees in his head the cave chart Painter had shown him concluding that ice has been melting since fire has been around. Noticing no pests flying around him for a change, Algor recounts the statistic Neander, the moth expert, cited – 40% reduction in bat population caused by fire killing the moths. According to the experts he met, Algor learned that fire is bad, and ice is good. He knows that fire melts ice, but doesn't fire keep people warm? Uncertain, Algor heads toward home, postulating all the clues he had gathered.

While walking thoughtfully, Algor admires the wood's inhabitants. With all its grandeur, Algor doesn't see a man lying under a brightly colored bush until he hears his meek voice cry out for help. It is an old man by himself, tired and alone. Algor is shocked to see this man's plight.

The shocked Algor blurts out, "You alright?"

The old man, barely able to stand, rises and mutters, "Me cold and tired."

Amazed, Algor asks, "Where is your tribe?"

"Left," moaned the old man.

Algor, still not understanding the situation, demands to know what happened to his tribe, "what do you mean left."

The old man, regaining his strength, reports, "Tribe all went south. Me was in charge of fire for tribe, but me cold and tired."

Algor instinctively asks the obvious, "Why not build a fire?"

The old man explains, "Entire tribe like pretty trees. No longer allow me to cut for firewood. Threw ax into river."

Algor's mind starts to put everything together.

"Fire eats trees. Fire burns things up. Fire leaves messy ash. Fire melts ice."

Anxiously, Algor asks another question of the old man," When stop using fire, did ice return?"

"So much ice, too cold to stay," recounts the old man.

Upon hearing this, Algor reaches his conclusion. The riddle is solved. The mission is over!

Consensus Is Reached

Algor runs with all his might toward home. He has a clear understanding and is excited to tell all the people, and then they would know what kind of leader he could be. Algor makes it home at last, still panting and tired, but smiling, because he has accomplished his mission. All the people gather in great anticipation to greet and hug him, asking what he has learned.

Soon Bilk gets wind that Algor is back and comes over to him. Bilk hugs Algor and sees his smiling face, then motions him to sit at his throne and tell everyone the story of his adventure.

With everyone huddled around the throne, Bilk looks to the anxious crowd and says, "Me Bilk, leader of all united cavemen and cavewomen".
Critically Algor replies, "Yes, we know, we know."
Bilk inquires, "Did Algor discover why ice and snow melting?"
Over excited, Algor exclaims, "Yes, the cause of all the united tribe's problems are because of one thing."
Bursting with anticipation, the crowd screams, "What, what is it ALGOR?"
Taking a deep breath, Algor sounds the alarm, "its fire!"

Gasping the crowd is astonished as Algor begins to explain what happened along his quest for the truth. Algor tells the story of how fire is causing so much heartache and pain, concluding that it must be banned if the united tribes are to be saved. Bilk, internalizing, wonders how he is going to get the people to stop using fire without becoming unpopular.

Bilk bites his lower lip and begins to mandate, "I will call on all the tribal governors to support our blue ribbon panel that will study this issue and make their recommendations." The fur salesman in the back runs away muttering, "we gonna need more furs!"

Algor is not satisfied, and thinks to himself, "This is not leadership!" Right then and there, Algor decides to quit politics and make it his life's mission to stomp out fires wherever they burn. And so goes the story of Algor, ex-Veep of all the united cavemen.

Mankind Evolves

Furthermore, my studies reveal that near the end of "man's first dark age", at the furthest tribe from Algor and Bilk, Eda, a single mom, and her son are eating dinner in their modest cave.

Eda's son is a brilliant, articulate and curious lad known for his ability to invent.

"Mom, I'm cold. Can I build a fire?" he pleads.
Eda firmly states, "You know our leaders have told us it is too dangerous."
Unimpressed, he implores, "Oh, Mom. How does a little fire hurt the environment? The wind blows the smoke away, and the trees grow back for more firewood. It doesn't make sense!"
Eda persists, "No means no. Our leaders know best."
Undeterred, he continues, "Mom, the uncooked food hurts my stomach, and your stew tastes better when cooked in the fire."
Chilled by the draft and softened by her son's reasons, Eda's motherly instincts overtake her anxiety. "Okay, son, go build a fire."

An hour later, across the path, the neighbors notice a bright, shiny glow emitting from Eda's cave.

"Look, honey. Eda's son is playing with fire again."
"Hey, don't they know it will melt the ice?" worries the wife.
"Why is that a bad thing? It's been so damn cold around here since the ban," curses the husband.
"That Eda's son is as curious as a saber-tooth kitten," quips the wife.
She adds, "Did you know Eda's son asked me to invest in his new invention - a round circular thing he says will revolutionize travel?"
The husband shakes his head, "That Eda's son is such a dreamer."

The wife asks, "Do you think we should invest our surplus in Eda's son?"

The husband looks down and sadly admits, "I had to give our surplus to a 6 foot 5 Neanderthal named Iris. Iris told me that the united tribes needed the furs for all the freezing people. Iris claimed that there are people less fortunate than us and the leaders need to keep them happy."

The wife replies, "You mean those lazy panhandlers who cheer when politicians give speeches?"

Husband nods, "Having to stand around and listen to politicians all day does make them less fortunate."

With a sigh, the wife replies, "At least we know now why it's cold all the time."

Research Conclusion

My research went on to discover that this tribe rejected the ban on fire and began to flourish. Eventually, they spread to multiple continents and adapted to different climates. Artifacts show this tribe was highly advanced, with the longest life expectancy and by far, had the highest standard of living. I dubbed these people – *"Conservators Capitalists."*

What happened to Algor and the united tribe is unknown as there is little evidence or artifacts to attribute to their existence after man's first Dark Age ended. We can only speculate that without harnessing fire, the hardships were too much for this species to survive.

Chapter 2 Pebbles in the Ocean

The goal of this book is to disprove the latest global warming doomsday scenarios with pure, unadulterated, logical thinking, rational reasoning and a heavy dose of common sense, which unfortunately appears to be waning these days. With my help and many common sense perspectives on climate changes, you will be able to stave off the bombardment of doom and gloom predictions hurled your way by a variety of liberal activists.

I promise to do my best to avoid manipulated reports, pointless statistics, bias surveys and hapless predictions based on tree rings and will try to get to the core of the issue. Straight talk, lined with a pinch of sarcasm, and a heaping helping of logic will provide all the ammo needed to combat the growing army of global warming fear-mongers.

For example, the next time an alarmists tells you the east coast will be under water by 2050, ask him what the weather will be ten days from now... They could not say with any certainty, right? Be assured that if experts cannot accurately predict the weather ten days into the future; forget about the dire predictions forty years from now!

These and many more retorts will comfort you and prepare you for the onslaught of media bias, whacko environmentalists and liberal socialist fear mongering. Their deluge of disinformation is intended to scare, slant and mislead you until you give in to their fascist demands.

This new wave of environmental activists supported by leftist scientists and liberal politicians is more determined than ever to convince you that we are destroying the Earth at a rapid pace. I have dubbed them "Global Alarmists". Whatever their motivations, we have to stop their unwarranted fear mongering. It is time for reason and common sense to prevail over scare tactics and doomsday predictions.

The advent and subsequent popularity of Sport Utility Vehicles (SUVs) has been under attack by the global warming decriers for more than a decade now. I equate automobiles and our atmosphere to pebbles and the ocean. How many pebbles would it take to affect the ocean? Too many to count, even with an environmentally friendly iPad. The fact is the emissions of automobiles on the planet compared to the Earth's atmosphere are equivalent to a bucket of pebbles thrown into the Pacific Ocean.

Hang in there, greenies. The way the world is sucking up oil, it won't be too long before we all are forced into a new mode of transportation. My guess is it will entail a more eco-friendly fuel. Perhaps hydrogen-powered cars will tote us around in the near future. I just hope we have a factory left to build them here in the good 'ole U S of A.

One sobering fact that got my mind straight on this whole global warming theory was the impact volcanic activities have on the global environment. A single volcanic eruption, like Mount St. Helen's in the 1980's, spewed more greenhouse emissions than all the cars ever built in the United States! We are talking about just one volcano; there are hundreds of active volcanoes at any given time around the world. Say this exploding volcano's name ten times fast: Eyjafjallajokul. Thanks Iceland, you are now the foremost polluter on the planet!

The Earth has been dealing with volcanic eruptions since day two... day one they weren't up and running yet. Seriously, to assert that the average automobile, which gets a few gallons per mile less than a hybrid, is causing global destruction is absurd. More to this point later - keep reading.

Global Warming or Climate Change?

The crux of the issue is, "Who controls the climate?" -- That is, people with a few million automobiles (though even the largest truck cannot be seen from space) or Mother Nature, who has been in the driver's seat for billions of years? Let's review a little history; scientists say man has been on the planet for a few *million* years, but only invented ways to emit Carbon Dioxide in mass quantities in the last hundred years or so. On the other side, we have the Sun and the Earth, which has been around for *billions* of years give or take a few months. I'd argue, to this point our planet has evolved quite nicely. If you don't believe me, go visit a nature park, Hawaii or Italy for that matter.

Historical facts have proven that the Earth has been through millions of devastating earthquakes, uncountable volcano eruptions spewing gases and ash into the atmosphere, mass flooding that has drowned everything under its flow, Category 7 hurricanes, tsunamis, meteor showers, the ice glaciers stampede and back again. So, based on Al Gore's scary movie of a 1 or 2 degree temperature increase, we are supposed to believe the end is near! I got my faith in this planet to outlive Al Gore and his cronies' doomsday theory.

If the planet is being forced to warm up by no cause of its own (global warming experts consider the Sun as a constant variable), then that is a different matter than if the planet is warming up on its own accord. The liberal environmental scientists want us to believe man is wreaking havoc on the planet and not vice versa. Remember, no one will go along with their plans to curb our liberties and progress unless global warming is man-made. We still have no control over Mother Nature, so it would be futile to propose taxing and regulating her.

To believe mankind can *control* climate changes is as arrogant as the notion we can even cause the global climate to change in the first place. Perhaps it is important to quantify just how large the planet and its atmosphere truly are. The Earth has a circumference of approximately 24,900 miles. The atmosphere has a vertical height of about 100 km (62 miles or 328,000 ft), creating a mass of 5×10^{18} kg. In short, it is gigantic.

What's interesting about our atmosphere (the center of attention in the global warming debate) is that it is 78% nitrogen and 21% oxygen, with trace amounts of water vapor, carbon dioxide and other gaseous molecules. Note the word "trace" and what follows it. If you listen to Al Gore Jr. enough (sic) you'd swear the atmosphere was mostly made up of greenhouse gases – It's not!

Trace molecules within the atmosphere do serve to capture thermal energy emitted from the ground, thereby raising the average temperature. Carbon dioxide, water vapor, methane and ozone are the primary "trace" greenhouse gases in the Earth's atmosphere. Without this heat-retention effect, the average surface temperature would be -18 °C or about 0.4 °F, and life would likely not exist. Thank god for the trace amounts of gases in our atmosphere. Personally, I hate when the temperature drops below 32 °F (that's 0 °C for you Europeans)

The global alarmists' biggest scare tactic is calculating (falsely) that the ocean levels will rise to some god-awful height, drowning everyone sunbathing on a beach. This scenario only flies if you melt just about all the land ice on the planet. Lucky for us, the annual average temperatures remain well below freezing in the main glacier areas. If you cherry pick your data, it is easy to sway a conclusion one way or another, so be careful putting your stock into any worse case analysis.

This is the main reason I do not attempt to impress you with foreign sounding names of renowned scientists and their infamous scientific studies or Emails from the experts at the Climate Research Unit. On the contrary, I prefer to regale you with clear-cut reasoning with a little sarcasm sprinkled in. I hope you enjoy reading this book as much as I did writing it… Oh yes, and the "…" all over the place are my signatory punctuation. They imply more thinking may be required…

Natural Selection

It is important to point out that some ecosystems struggle, while others flourish. Evolution tends to be slow (take liberals for example). So you must look at weather related issues globally and over very long periods of time. Unfortunately, in the scheme of things, comparatively we only have viable weather data for a short period of time, and even that is spotty. Tree rings are far from definitive proof of temperature change. In fact, they are only definitive proof the tree existed.

The Earth is not stagnant. Earth has ways to filter itself, regenerate, adapt and be self-correcting. Maybe not as quick as a drive-thru, but yes, the Earth's ecosystems do *self-correct*, or they die out. Often, this is a slow process (longer than 100 years). So, in today's fast-food world, it does not faze me that some people cannot think in such long-term cycles.

Earth has the capacity to evolve and adapt. When I talk about the "Earth" in this book, I am referring to the Earth's ecosystems (a natural unit consisting of all plants, animals and micro-organisms in an area functioning together with all of the non-living physical factors of the environment). Areas that are arid have plants and animals that benefit from that particular condition, just as wetlands offer benefits to other types of plants and animals. The point is that no one climate is optimal for all plants and animals. Each species has adapted to its surroundings and will continue to do so.

Before man-made emissions, plants and animals adapting to surrounding environments have made the ecosystems strong. To suggest that it is somehow in peril from a minor temperature increase is ludicrous. That would be like worrying that the Roman Coliseum is in danger from spit wads.

The Sahara was not always a desert; few beaches were sandy, and the Grand Canyon was not always a canyon. But over time, these areas formed into what they are today. All the plants and animals evolved with these environments and will continue to do so, even if the Sahara Desert becomes a wetland again, or the Grand Canyon fills in and becomes flat! Plants, animals and even man will continue to adapt.

I would argue that the Earth is better off with calamities. It makes the whole system stronger, because plants and animals are forced to adapt. Mutation is one mechanism in which living organisms change and eventually evolve. Some changes increase the chances for a species to survive, while others hinder its survival rates. The ones with the best genealogy propagate. I heard this in school once called *natural selection* or something like that…Google it.

Living organisms that don't adapt die out (a phenomenon that has been happening since species existed), while others that do adapt take their place. This all happens for a reason - to make the whole ecosystem stronger. The world as we know it has not survived millions of years by being fragile!

Been There, Done That

Let's look at what we do know about our planet's history: this planet started out as a big ball of cosmic dust that heated up to molten rock and after millions of years, it cooled off enough to support life. This was the very first instance of global warming and global cooling, we just didn't have any liberals at the time to complain about it.

Unrepentantly, the planet over cooled, creating what we now call the *Pleistocene Epoch* (a series of ice ages). Then, miraculously, without any help from FEMA, the planet corrected itself and cranked up the thermostat (note, this is still prehistoric time). One could argue that was a catastrophic event, perhaps slow to progress, but nonetheless, initially devastating to the environment at the time. Doesn't this prove at least the possibility that the Earth can alter its own climate without the help of man?

With no man-made smokestacks or trucks to speak of back then (i.e., Prehistoric Era), there are no humans to blame for the multiple ice ages, only Mother Nature (unless you propose the dinosaurs could have caused them). Somehow, someway, the root cause of today's weather calamities is rarely blamed on Mother Nature. It is said by the global alarmists that only man-made emissions are causing the problems.

By the way, when did Mother Nature skip town and let the evil corporations take over the weather anyway?

The bottom line is that massive weather changes have been occurring long before the combustion engine was ever conceived. Is the planet going through another catastrophic phase? Perhaps, but at a half a degree a century, I am not scared. Put another way, what reason does anyone have to suggest the Earth will never go through another cycle of change? In fact, I'll bet that in the next million years or so another climate change will occur, regardless of any human activity. I guess my great grandchildren will have to collect on that one…

When I was a young student, I heard a respected scientist claim that the Sun will burn out and the planet will die. I remember the shock and awe I felt just pondering the pronounced Armageddon. How horrific it would be to be alive at the time the world was ending (wait for it, wait for it). Until a few moments later, the scientist stated that this event wouldn't happen for another ten billion years. "Phew," I thought to myself, "that was a close one."

Why hasn't Al Gore, Jr. or his minions started a campaign to save the sun? If and when the liberals do enact their plan to save the sunshine, I am certain it will include some very creative regulations and carefully targeted tax hikes. Perhaps by then, liberals will have evolved enough to realize government cannot solve every problem. I like to ponder the possibilities no matter how slim the odds…

Here is my doomsday prediction: When the Sun burns out, at no fault of ours, life, as we know it, will end. No government regulations, no liberal policies, no blaming the conservatives will stop this from happening. Nothing is more certain than death and taxes, and nothing is more resilient than Earth.

Old News

A news report surfaced recently that the winter weather this year would be warmer than last year in the Northeast, except Maine. Seriously, they can predict Vermont and Massachusetts will be warmer, but a few miles north and east will not? It's amazing how precise the predictions are which are being asserted by experts and how they can be reported as fact by journalists. I guess a year from now that prediction will become old news. I doubt anyone will hold the weather experts or journalists to their accuracies or inaccuracies...

If weathermen really wanted to be more accurate, they would tell us what the weather was like last week instead of next week, giving them a much better chance of being right. However, forecasters will be forecasters, and they continue to guess at the weather with immunity. No one is safe from the forecaster's prognostications. Has anyone ever documented the accuracy (or inaccuracy for that matter) of these TV experts? No one seems to care how far off the actual weather these "experts" are, but I do!

One year a local news station came up with a unique promotion called "accu-weather" or something like that. They promised to give $100 to charity if their daily temperature prediction was off by more than +/-5 degrees. Needless to say, they could not afford to keep the promotion going, and I think they fired the clown who came up with the idea. He now works for Hadley CRU.

More good news: the reports claiming that we are experiencing higher temperatures are leaving out some very interesting facts. After further review (sports terminology), the majority of the purported warming is actually occurring at night and during the winters. So, again we have to put into perspective the *ravages* of milder winters (less accidents and heating oil) and the *dangers* of warmer nights means one less blanket for me.

The Protocol Sun

One other thing that befuddles me about the global warming scientists (more like theorists) is that they treat the Sun as a constant in their global weather models. I'm sorry, the Sun is there for us every morning when we wake up, but it is not perfectly constant. Anyone smarter than a 5th grader could tell you that the Sun is a huge perpetual mega explosion (Answer C: fusion). It's astonishing the amount of energy our Sun cranks out every second of every day and has so for billions of years.

Isn't it amazing that the global models used to predict the future temperature don't take into consideration the sun's solar flares? With hundreds of variables factored in, it is amazing that the single biggest factor (i.e., our heat source) is treated as a constant. If you think the Sun is a perfect **11,000 °F** all the time, I am sorry to inform you that it is not. Like everyone else, it has its good days and its not so good days.

Our star, the Sun, is like a million atomic bombs going off every minute, and by the way, the phenomenon called solar flares does cause fluctuations in the sun's energy output. Some solar flares have been so strong that they have affected communications here on Earth. If that is possible, then why isn't it possible that increased solar flares have caused some of the real or perceived global warming...? Sorry, I am not ready to concede man-made climate change after reading East Anglia's CRU Emails.

Since the Sun plays the predominant role in determining our Earth's temperature, it stands to reason the planet's temperature fluctuations are more likely due to variances with the sun. You can put your hands over your eyes and claim it is dark, or you can open your eyes and see the shades have been drawn. Don't get snookered by flaming liberal fear mongers who need you to be afraid before you will hand your car keys over to them.

It is true that in any controlled chaos there are fluctuations; nothing to panic about, but the Sun does have *lite* variations from time to time: sunspots. Like, solar flares they are naturally occurring fluctuations that do affect Earth (e.g., airwave transmissions, temperature) and yes, its overall climate.

Incredibly, these known, perhaps unpredictable, phenomena are not accounted for in the most basic global warming models. How could you leave out the most important climate-affecting variable of all, unless you were not interested in ascertaining the truth!?

Benefits of Global Warming

You never hear about the benefits of global warming in the news do you? Of course not! That would negate the fear-mongering effects of the award-winning movie "An Inconvenient Truth", staring an award winning actor – Albert Gore, Jr. Never underestimate the power of the big screen to influence the masses. Now that there is money to be had, beneficiaries are piling in hoping to get a piece of the pie (mostly American pie). Not only am I not jumping on the band wagon, I am throwing myself in front of the horse in hopes of slowing it down long enough for *cooler* heads to prevail (all puns in this book are intentional.)

For you Dave Letterman fans, here are the top 10 reasons *not to be scared of global warming*:

1   Longer harvests equals more food (not everyone is obese)
2   Just imagine all the heating oil we will save (sorry big oil)
3   Migrating birds won't have to travel so far (score one for the animal lovers)
4   Less iceberg related collisions (no sequel to The Titanic)
5   Longer swimming seasons for outdoor pool owners (more members for the polar bear club)
6   Longer BBQ season (who do you think lobbied for longer daylight savings?)
7   More vegetation means better looking flowers (girls get tired of the same roses all the time)
8   Women can wear bikinis and short shorts longer (score one for the men)
9   Fewer snow days for schools equates to smarter kids (watch the show - *are you smarter than a 5th grader?*)
10 Drum roll please... It may get so warm that people will no longer celebrate Christmas. Then the liberals would get their ultimate goal - all religion out of the public life. Now if it weren't for Easter -- oh wait, we can just call it spring break instead

* Keep reading to see my personal favorites...

Every single one of the top ten has a bit of truth no matter how jovial they sound. Seriously though, we can think of many significant advantages of a warmer climate that negate the dire predictions of death and destruction from the global alarmists. The human body likes temperatures around 72 degrees Fahrenheit, right? Currently, the US average is a bit below that, so as the temperature rises, it actually approaches the human ideal temperature. Additionally, where I live in the Northeast, wintery road conditions account for millions of injuries and billions in property damages; warmer temperatures would certainly mitigate at least some of the needless deaths and damages for passengers of all ages. Save the women and children, warm up the planet!

Back when I was in school, I was taught that plants need carbon dioxide to live and that pure oxygen would kill you. So why isn't oxygen the pollutant and CO_2 just left alone? Everything in totality must be considered or issues become skewed and pointless to argue.

"Nothing astonishes men so much as common sense and plain dealing."
Ralph Waldo Emerson

Chapter 3   Is It All Bad?

If I wrote this book thirty years ago, it would be arguing why man is not causing the temperature to get colder. For those of us old enough to remember or astute enough to know, the environmental whackos of yesteryears were clamoring about the impending ice age. See the Dec 24th, 1979 issue of Time magazine.

Since that "melodrama" was a flop at the proverbial box office then, the doomsayers went back to the drawing board, but this time scurried to the other end of the spectrum. With their new global warming theory and fudged data models in hand, once again they are predictably spewing alarmist condemnation yet again.

One of the knocks against Nostradamus's prolific predictions is that he made so many of them, (most so general) that some of them were bound to come true - law of averages. The same can be said for the environmental whackos trying once again to frighten us into cowering like a herd of sheep so they can corral us into their fleecing machines. Remarkably, reason is still alive, and I will continue to tap into the common sense and rationale of all people concerned with the rogue global warming activists.

"One man's junk is another man's fortune." I'm not sure who invented that quote, but I'm guessing it was from a capitalist. More importantly, it demonstrates that it's not all bad for the planet. I am certain there are many residents of Buffalo, New York, or even Siberia, down on their knees thanking God or, in the case of the Russians, thanking the Kremlin for warmer temperatures.

Thinking on the bright side is a favorite hobby of mine (you may not know that from all the sarcasm). Every event has different perspectives. There are many clichés to that effect like, "Is the glass half empty or half full?" What good can come out of a change in venue is what's important. Since change is inevitable, let's all try to embrace it. It really beats hopelessness and fear. I know.

So, before I finish ripping Al Gore, Jr. and his cronies' theories to shreds, I jotted down some of my personal favorite benefits of global warming, just to list a few:

- Less harvest lost to pesky frosts (some of us are still hungry)
- Warmer temps where it is unbearably cold (Siberia, Buffalo NY, etc.)
- The next ice age may be prevented from occurring (go Sun)
- Mountains increase in height due to melting glaciers, becoming higher as they rebound against the missing weight of the ice (better skiing and/or snowboarding)
- Boundary disputes between countries over low-lying islands will disappear (peace on Earth at last)

The doomsday profiteers will either attempt to discredit me (good luck, since I used logic as my basis) or perhaps ignore me hoping too few people read my book. Either way, they cannot argue the other side without plunging holes in their own theory.

"Freedom is never free."
Unknown Patriot

Oh Happy Days

As promised, I am trying not to throw useless, bias-riddled data and statistics at you, clogging your mind from pure logic and common sense. So bear with me as I assist your own experiences and reasoning to fill in the murky areas. Even if you live in a cave (without fire) like some liberals want us to go back to, you must be somewhat aware of how vast and sophisticated the global climate is. Your instincts have probably already told you that we can survive a one or even two degree hike in average temperatures. They were right, read on...

Fact - millions of people worldwide heat their homes and businesses. This costs money, time and resources. Some whackos even claim this is melting Greenland's glaciers. Well, logic dictates that warmer temperatures would reduce everyone's energy burden tremendously. I mean with fewer furnaces running, wouldn't that reduce greenhouse emissions?

I am sure someone could calculate the approximate dollar savings each degree of warmth would provide every household. Just image all the heating oil we would save. However, it just doesn't matter that overall the new found warmth would be a good thing; the global alarmists want to ignore that.

I live in the northeast and see first-hand the misery and cost of living, working and driving in the ice and snow. Every year millions of people are daunted with the trek through blizzard conditions. Again, somewhere on the internet is the exact cost in terms of life, limb and property losses due to icy cold temperatures (I have personally lived through several severe ice storms.) Oh, the happy days, when the temperature hovers above 32 degrees (a.k.a. 0 Celsius). With all the money we would save from a warmer climate, everyone could afford a swimming pool and a new barbeque grill.

Returning to a capitalist outlook, think about all the businesses that would benefit from a few more degrees of heat - the travel industry, the golf industry (one of my favorites), and perhaps, even the flower industry - all of which thrive on milder climates. Basically, for every business that falters from milder temperatures, I can list two that will benefit. So, either focus on the negative (more sunburns) or focus on the positive (thriving sunscreen, hats and sunglass industries). I prefer to enjoy the sunshine versus complain about it...

Back to nature - "But the birds and the plants can't profit from the impeding sunshine," say the doomsayers. Perhaps geese and pine trees cannot benefit monetarily, but they do benefit from milder temps. Most animals spend much of their energy on staying warm (probably because they live outdoors). Trees love the Sun and in return they give us shade.

Wild animals have to spend much of their time preparing for winter and sheltering themselves from the cold. Just think how much more free time bears would have if they did not have to spend all summer fattening up for hibernation? The examples are boundless. However, the bottom line is that milder temperatures promote longer growing seasons, which equates to more food. More food means more time to frolic in the sun. See how logic works across the board here? We have an amazing planet, don't you agree?

"Common sense is in spite of, not as the result of education. "
Victor Hugo

Did You Notice?

Since most, but not all, scientists seem to concur that the average mean temperature of Earth has increased over the past hundred years or so, I would be a contrarian to argue against them. So I will concede this to further the acquisition of the real truth. What I will not relent on is why the temperature is changing AND whether it is a bad thing! I will continue to discuss why a minor temperature increase is a good thing for us. But, first we must probe the question, if the temperature is changing, why is this occurring and why should we not panic?

We are being told that the global temperature has risen a half a degree over the last fifty years. So what! The temperature where I live (the Northeast) can change thirty degrees in one day! If you were sitting in your comfortable living room and Al Gore Jr. snuck in and turned your thermostat up a half-degree, would you notice? Would the paint start peeling off the walls? Would all the dishwater in the sink evaporate? Would the ice melt in your freezer? Of course not, because your household is not that fragile and neither is this planet.

Historically speaking, since the Earth has been through a tad more turbulences than a temperature change of a degree or two, I submit that it can and will survive a lot more than what we are throwing at it today. Remember the average people are only throwing pebbles into the proverbial ocean. Even though some people with Boeing 747s like Al Gore are throwing stones, the oceans are still waving.

As I just alluded to again, nowhere on the planet is the temperature stagnant. Nowhere is it perfect all the time. It fluctuates, as most of us know. If the temperature were a constant 72 degrees everywhere in the world day and night, night and day, then and only then would I worry about a change in the temperature!

Since the beginning of time, the global environment has dealt with constant change, even traumatically destructive events. I have not even gotten to the fact it goes through seasonal changes every few months thanks to the tilted axis of the Earth and that ever constant Sun. The point is, simply, <u>do not underestimate the prowess of nature to handle climate variations</u>, even drastic ones, and to be able to overcome them -- it always has, and until the Sun dies out in approximately ten billion years, it always will...

How significant is .25 or .5 degrees in the scope of things really? If it averages 19 degrees in the Antarctic, then .5 degree increase still keeps the temperature below freezing. Only where the temperature is 31.76 degrees the ice may melt. Experimenting, I changed my freezer setting from coldest to cold a few days ago, and amazingly the ice cream is still frozen...

If the warming continues at the torrid pace of .5 degrees per century, when can we expect direct catastrophic consequences? Catastrophes will continue because of either man-made events (terrorist's dirty bomb going off in a mall) or naturally occurring events like earthquakes. In either case, the temperature will not be the cause. So, keep your insurance up to date, but do not buy into the new Y2K hoax being perpetrated by the latest environmental hucksters.

We are told by the global alarmists that we will have more frequent and more intense heat waves. What is the definition of a heat wave? Does it mean it's hot out for a long period of time? Like summer? Weren't they called the *dog days of summer* by our grandparents? Now summer is something to be feared? Lucky for us the Sun goes down every evening like clockwork.

What if the average temperature during summertime in Nevada is 92° F? Does that mean if it reaches 102° F for a couple of days it counts as a heat wave? How about Juno, Alaska where the average summer temperature is about 56° F? Does this mean a prolonged period of 66 °F constitutes a heat wave? I doubt the Alaskans will have heat strokes, as I hear they are a hardy bunch up there.

The point is that it is all relative. The temperatures fluctuate based on many variables, including the seasons. A heat wave in the winter rarely draws much attention, only when it is over 100 °F in a city does it cause newsflashes. In fact, I heard the Siberians have been buying SUVs in droves since Al Gore Jr. convinced them it will increase their temperature if they drove them around.

Run for the Hills!

I live near one of the Great Lakes and every year there is a hullabaloo about the lake's water level - some years it is too low, others, it is too high. Regardless, no one's house has fallen into the lake. Amazingly, no cataclysmic event occurs year in and year out. Why? According to global alarmists, we should be experiencing coastal flooding any day now. Obviously, the lake is not the ocean, but the principle I am asserting still applies – water levels fluctuate.

The lake level changes, like many ocean fluctuations, are usually within the environment's tolerances (for exceptions, see hurricanes, tsunamis and tidal waves). As I asserted earlier, the environment has and will continue to endure changes! If you don't believe me, then please don't live near or even visit a coastal area?

Ocean levels raise and lower everyday; it's called tides! No large body of water is always calm nor does it remain the exact same level forever. Just as land temperatures swing up and down, so do the ebb and flow of large bodies of water around the globe. That is why beaches are usually long and wide with sand dunes usually existing just inland. Many waterfronts have cliffs or other natural barriers to keep the water at bay, literally. The point is every coastal area I know of handles water level changes on a daily basis. Remarkable how nature thought ahead...

Notice next time you are at a beach (or any waterfront for that matter), you will inevitably see a barrier just inland, perhaps even a rocky cliff that prevents waves and rising waters from overcoming the inland areas. This is all nature's design to handle fluctuations in water levels that naturally occur *all the time* (thanks, moon).

Furthermore, the evolutionary tide is changing as man has figured out ways to minimize and control water levels in certain areas; these inventions include dams, levees and break walls, etc. This is a testament to man's ability to adapt and overcome natural forces that work against him. I could list a thousand more, but I hope you got my drift...

Just like the temperature changes naturally, water levels also change naturally and sometimes violently, and not because the waves are mad at us for letting our kids pee in the water while swimming (kids, what are you going to do?). No, they just swell sometimes like liberals do when they are losing an argument. Blame the wind, not the retracting glaciers.

It amazes me how the environment and sometimes man has adapted to handle natures many furies. Look how the Nile flooding actually allowed farmers to cultivate the land (look up *silt* in Wikipedia). Visit the dikes in Holland or see the terraced rice paddies on the mountainside of Japan. To sit here and discuss how teeny alterations in the climate can devastate the planet is small minded, arrogant, fear mongering or just plain disingenuous.

I grew up three miles from the Great Lake Ontario - also the same distance from one of our few nuclear power plants. (Yes, my children have ten fingers and ten toes.) Furthermore, I paid attention in school when learning about the *undisputed* ice age that carved out all the Great Lakes. So, if the fragile Earth with all its diverse life has the capacity to survive changes from tropical temperatures to an invasion of mountainous glaciers and back to moderate, relatively speaking, temperatures again, why should we duck and cover over a degree or two?

Frankly, I would sound the alarm bells and evacuate the schools myself if evidence suggested the Earth is **not** changing, since that has never happened! It is imperative people put the global warming claims into historical perspective. Fortunately, common sense and rational thinking can win the day and rescue hope back from the global alarmist's death and destruction outlook.

This just in: I called the Island of Manhattan and it reports that it is still comfortably dry, even with the predicted ocean level changes. The next New Year's Eve celebration is safe for now.

Plants Aren't Stupid

Don't more plants grow where it is warmer? I am certain there is more vegetation in milder climates. So doesn't it make sense that a longer growing season in the far north or south would reap more fruit? Watermelon is my favorite, but it will not grow in New York. This is not because of the high taxes here, but because of the lack of sunshine. Perhaps my grandchildren will be able to grow watermelon here someday. Sometimes you just have to look on the bright side. It must be depressing to always live in fear. How about it Al? When is the last time you said something cheery to someone other than your accountant?

So, on the positive side, warmer temperatures equate to longer harvest seasons, which equals more food. That is a good thing, as our population seems to increase every census year (legally and illegally). What about the impending floods? Not to worry.

Someone invented the break wall. I have seen them, and they work great when the waves get too tall. Seriously, every coastal area on the planet is capable of handling fluctuations in waves and tides, etc. Nowhere is a large body of water always calm. On the contrary, it is usually smashing pebbles into sand.

Please think back to grade school and recall just what this Earth has been through over its lifetime. This planet has survived earthquakes, volcanoes, floods, too cold, too hot, too dry, too something. You name it - Earth's been through it. The question is, "Do you believe in Al Gore, Jr.'s bleak future or the proven history of our own planet to survive?" This planet is completely amazing! The ability to endure hardship, sustain life, and provide balance is, quite frankly, awe-inspiring.

Also, I am betting on our ability to be good stewards (something liberals do not believe conservatives can do). No, in their minds only liberals can conserve and protect the plants and animals by instituting laws and regulations. They believe conservatives all want to profit at the expense of the environment. They don't seem to believe that conservatives need clean air and water too.

Liberals think that no conservative has visited a national park or just enjoyed the beauty that is nature. It is beyond liberals to think we all benefit from a more pristine environment. The truth is conservatives would rather adopt real solutions to problems rather than roll back progress toward the Stone Age. We embrace conservation; we just don't want to take all the fun out of it...

Conservatives believe that we humans can overcome issues we face without the government mandating implausible solutions. We have faith in knowing we can save the planet (if we had to) while continuing to improve our standard of living. Think of it this way: If we could stop Darth Vader's Death Star and make a profit at the same time, what is wrong with that?

We do all have the inherent responsibility to manage our resources and bestow that ability to other peoples of the world. Perhaps Al and I agree on this point, although we just differ on how to go about it. His solution entails higher taxes and stifling regulations. Conservative solutions promote free market incentives and technology. The difference could not be more pragmatic.

"Stay off the grass!"
Grumpy Liberal

Desert Life

You might think that animal and plant life in the desert would be sparse. The high temperatures and lack of water do keep many plants and animals from living in this area, but life does thrive in the desert. Amazingly plants and animals have **adapted** to allow them to live in these extreme conditions.

Many animals conserve energy by sleeping in underground burrows during the hot turbulent day and hunt for food in the cool night. Insects and spiders that live in the desert have thick, hard bodies that help maintain their body temperatures. Reptiles have scales, which keep in moisture. The list of sophisticated attributes, like the human thumb, prove that life has evolved and will continue to do so, even if it seems the liberals need some catching up.

Plants in the desert also have adapted to the harsh environment. Many plants grow in just the few weeks during the rainy season. They produce seeds quickly and then they die. Other plants have shallow root systems that soak up water quickly after a rainstorm. Still other plants have very deep roots that can reach water deep within the ground. The ability to adapt is the hallmark of all living creatures, and they are testaments to what is possible today.

I remember learning as a young student about how the 30+ temperature swings in the Death Valley area (in both California and Nevada) occur from day to night. It was mind boggling that anything could survive these incredible temperature swings, not only seasonally, but also on a daily basis! Ironically, to global alarmists, they do survive. I say ironically because according to the Al Gore's disciples, everything is too fragile.

Granted, only the hardiest of creatures can survive the desert, and they have had many hardships to adapt to, but they do it! So, whenever I hear another fear monger proclaim the end is near because the temperature went up a ½ degree this decade, I fall back on logic, experience, and instincts that tell me something's fishy here. No way can this Earth cave in to such an insignificant, slow advancing alteration. Not after comprehending historically what this Earth has been through and what is evident on a daily basis just outside our windows...

Another scare tactic of the left is to predict that a million species may be driven to extinction by 2050 - a blatantly absurd statement conjoined to the hip of global alarmists and environmental whackos. The truth is that many species do die out every year, just like many glaciers melt every summer. What they omit from the equation (making the comment less scary) is the fact that many new species are discovered every year, and many new glaciers freeze every winter.

If nature did not balance itself, we would not have the diversity we see today. Furthermore, to tout death rates without factoring in birth rates in an attempt to skew the statistics is misleading at best, or fraudulent at its worst. I have never known a politician to skew a statistic, have you?

It is important to put everything in context here. The struggle for survival has been ongoing since the planet became habitable, and life began to flourish. Nature, or God if you prefer, has set up a sophisticated and marvelous system of micro and macro ecosystems. As big as a continent and as little as a puddle, life weathers the environment, battles other competing species and instinctively strives to multiply.

However nature came into being, the fact remains that the system favors the strong. The plants and animals best able to adapt and to endure cycles of calamity thrive, while others die out. No one is personally to blame, that is just how it is, regardless of your ideology.

"Everybody gets so much information all day long that they lose their common sense. "
Gertrude Stein

Contrarian

Call me a realist, but regardless of labels, imagine this: if the experts came back and told us the average mean temperature everywhere in the world will be exactly the same a hundred years from now, I would sell everything and run for the hills. Or better yet, I would buy an island and live like a survivor show contestant…

The day or century this ever-evolving Earth stops changing, we are all dead or at least on our last leg. The time to panic is when the Earth stops adapting and reshaping itself along with every living organism. Until then, I am quite confident that we as humans and 99% of the plants and animals will continue to endure much greater hardships than a fraction of a degree more warmth brings.

According to the global alarmists, the temperature has been rising for the last hundred years. Let's just say that is the case for argument's sake. Taking that into consideration, why have we not seen the calamity from this so-called drastic man-made weather phenomenon? Sorry alarmists, Katrina was caused by the wind. If Katrina was caused by the global warming effect, then why haven't we experienced category 5 hurricanes every year or every month?

If a one or two degree temperature rise is so unsustainable and destructive, then why haven't large numbers of civilizations already perished? Last I checked no island has sunk under the sea (except for Atlantis, but that was a long time ago). No plant or animal species have gone extinct that would not have done so otherwise. Sorry, alarmists, you cannot blame every species of gnats that go extinct on global warming. In fact no catastrophe has yet to happen because of global warming.

Every storm or weather phenomenon that we have experienced in the last 50 years has happened somewhere before in the world and is well documented. So, if we are living in a cataclysmic period, why are there no <u>new</u> weather occurrences to indicate that we are living in the last days of planet Earth as we know it? The answer is because we are not, at least not due to a minor climate change.

Nothing has happened of late that has not happened ten-fold throughout the pre SUV centuries. Why? Because a half-degree change in the mean average global temperature is not cataclysmic; it is a natural state of fluctuation. Like the ebb and flow of the ocean, global temperatures are not constant and never will be.

The scare tactics used by the global alarmists are old and transparent. I like to say they are full of hot air! It frankly surprises me that so many consider their doomsday scenarios might be true and are willing to fork over their dough or worse yet their sovereignty. I do have to admit one thing the Gore types have going for themselves - it is hard to disprove something that is not provable...

"Common sense and nature will do a lot to make the pilgrimage of life not too difficult. "
W. Somerset Maugham

Chapter 4   Albert Arnold Gore, Jr.

I like the reference Rush Limbaugh uses for **Albert Arnold** "Al" **Gore, Jr.** calling him *Algore*. I don't think Rush would mind if I borrow this alias because we see eye to eye on many issues, including Algore's ulterior motives. In any case, I prefer my own alias, as you will see throughout the book...

Ulterior Motives

Albert Gore, Jr.'s whole life revolves around the quest for power. He has dedicated his life to achieving it. From son of a senator through his ascent to Bill Clinton's Vice President, he has demonstrated an egomaniacal tendency. I am certain that failing to win the coveted presidency in 2000 AD (post Bilk presidency) had left a huge void and knowing now the chances of ever running again are next to nil, must be fuelling the fire of his discontent.

What better way to regain some self-respect and a little limelight then launch a campaign with no voters. Who would have imagined that just by wearing a doomsday sandwich board and making a *mockumentary* could one person remake himself as the last great messenger? I say, "last" because according to Gore, life as we know it will be gone within the next dozen years or so.

Gore and those like him are crazy if they truly believe that our lawn mowers can effectively change our climate. Every day we hear how some modern inventions (other than the Internet) are causing dangerous levels of carbon emissions. It takes a lot of arrogance to pronounce the end of the world without considering what the Earth has already been through, and it shows *ye of little faith* by not contemplating what this planet can survive.

This fact leads me to conclude that there are other factors in Albert Gore, Jr.'s pursuit as world savior. Since he will probably never let us in on the truth, we may have to deduce his motives from prior actions and statements. This won't be difficult, given that this politician has been consistent on important issues, revealing his true agenda: quest for power.

This race is most certainly easier than winning the Presidency as it is a one-issue campaign, and it has no term limits. With no Supreme Court to adjudicate against him, Gore is definitely smiling again because he knows he is back on top - for now, and the only opposition this time is reason.

Reason is something liberals revile, but somehow manage to overcome. Facts, logic, and results no longer get in the way of accomplishing their liberal agenda. Emotions rule the day; when enough people want to believe the mincemeat you are feeding them, it is easy to feel invincible. Albert Gore, Jr. is laughing again because he has snookered the unsuspecting, uninformed, irrational half-wits into believing his doomsday scenario.

When a failed VP of the United States makes it his life's mission to become the pied piper of the environmentalist movement, we should all take a step back and carefully analyze his motives. Do some liberal politicians have egos? What do egocentrics do when they fail to reach their dream of being President of the United States? Does this newfound fame make Al Gore Jr. relevant again? It doesn't take a genius to figure this out, just a thinker like yourself.

Albert Gore, Jr. is a power hungry, egotistical ideologue, who under the guise of "Planet Savior" will stop at nothing, retract nothing, nor feel any remorse for needlessly and selfishly fear mongering. He is the terminator Sarah Conner should fear the most: a human face on a robot hell bent on destroying our American way of life. Only the keenest of senses can sniff him out. Congratulations on finding this book.

Could this be a way for Albert Gore, Jr. to get back at the rightwing supporters who supposedly swift-boated him out of his lifelong dream? I say "supposedly" because it was actually the mean old Supreme Court Justices who snatched away his victory by ruling against him in 2000.

Wouldn't this be the greatest episode of "Punked" ever produced in history? I could see Gore getting Ashton Kutcher to come on tour with him to say, "You have just been punked!" Look who stands to lose the most -- the very people who denied him his coveted seat of power. Wouldn't that be the sweetest revenge Albert Gore Jr. could ever imagine?

Nobel Piece Prize

Congratulations, Al, on your *piece* prize (sic). I should also congratulate the Nobel committee for demonstrating how meaningless their award has become (this prize has nothing to do with peace anymore.). What has needlessly scarring the population over trace CO_2 emissions got to do with world peace anyway? If anything, he has generated more unrest (Last I checked is the opposite of peace).

58

What a joke. I never heard of a nominee actually lobbying for the prize before. Leave it to Gore to break new ground on this selfish maneuver. (See details of Al's trip to Switzerland prior to the selection.) When are people going to learn it is all about Al Gore Jr.? It always has been, but that is normal for egocentrics. You won your coveted prize, Al, now go home and turn down your thermostat!

You know Albert Gore Jr. is basking in his glory. I mean how many people can say they fooled so many people. His reign of terror will end just like Thomas Aquinas unraveled the flat Earth nonsense. So we thinkers begin to refill the grave the alarmists have begun to dig for us in order to bury ourselves. In their minds, the meek won't inherit the Earth; the elitist will.

Why don't they give people awards for real solutions? It seems the biggest complainers are getting all the attention. I know, it is the liberal mindset to complain, and it's not the actual results of their programs, it is their intentions we must focus on. Please! Is that what we have come to - honoring the whiners instead of the solvers?

Nobel Rolling in His Grave

Who is on this Nobel Peace Committee, and whom did Albert Gore Jr. beat out for this prestigious award? Inquiring minds want to know. Prior winners include Jimmy Carter Jr., former President of the United States of America, for his decades of effort to find peaceful solutions to international conflicts, to advance democracy and human rights, and to promote economic and social development. That effort worked out so well peace is still breaking out all over the Middle East. And I thought he won for inventing the misery index...

I will take back everything I say about the committee if they would select winners based on actual solutions to real world problems. Yes, Al Gore Jr. has been successful at scaring people into driving Priuses and switching to mercury lined fluorescent light bulbs. But to give accolades and cash to a *Henny Penny* "the sky is falling" con man is ridiculous.

The foundations for the prize were laid in 1895 when Alfred Nobel wrote his last will, leaving much of his wealth to the establishment of the Nobel Prize. Alfred Nobel was a scientist, inventor, entrepreneur, author and pacifist. He is rolling over in his grave, now that an ex-Veep, turned con man, won his prize and is using it to validate his claims about man-made climate changes.

Don't let this lottery win validate Albert Gore Jr.'s scam. It's like suggesting a mega power ball winner was successfully hedging his investment portfolio. Don't think politics are not involved. The selection committee is made up of human beings with biases and agendas of their own. What better way to validate their own liberal thinking than to select one of their own?

Do as I say, not as I do

Referring to Al Gore Jr., do you really believe the ex-veep is traversing the planet in his Boeing 747, peddling his doomsday scenario, out of his compassion to save us from ourselves? If you do, I'm sorry; you may be a little naive. I love to give people the benefit of the doubt, but I know the liberal track record too well. For example, consider the continuing transfer of trillions of dollars for the war on poverty (and you thought the war in Iraq was expensive; at least we are winning that war).

Public records reveal that as Gore lectures Americans on excessive consumption, he and his ex-wife Tipper lived in two properties: a 10,000-square-foot, 20-room, eight-bathroom home in Nashville, and a 4,000-square-foot home in Arlington, Va. (I heard he sold his third home in Carthage, TN because he bought a nicer home in Malibu, CA). For someone rallying the planet to pursue a path of extreme personal sacrifice, Gore requires little from himself.

Then there is the troubling matter of his energy use. In the Washington, D.C., area, utility companies offer wind energy as an alternative to traditional energy. In Nashville, similar programs exist. Utility customers must simply pay a few extra pennies per kilowatt-hour, and they can continue living their carbon-neutral lifestyles knowing that they are supporting wind energy. Plenty of businesses and institutions have signed up. Even the Bush administration used green energy for some federal office buildings, as are thousands of U.S. residents; but not Al Gore.

Al Gore Jr. is not alone in his hypocrisy. Democratic National Committee (DNC) Chairman Howard Dean has said, "Global warming is happening, and it threatens our very existence." The DNC website applauds the fact that Gore has "tried to move people to act." Yet, astoundingly, Gore's persuasive powers have failed to convince his own party; the Democrat party has not signed up to pay an additional two pennies a kilowatt-hour to go green. For that matter, neither has the Republican National Committee. Maybe our very existence isn't threatened.

Al Gore Jr. has held these doomsday views about the environment for some time. So why then didn't Gore dump his family's large stock holdings in Occidental (Oxy) Petroleum? As executor of his family's trust, over the years Gore has controlled hundreds of thousands of dollars in Oxy stock. Oxy has been mired in controversy over oil drilling in ecologically sensitive areas. Living carbon-neutral apparently doesn't mean living oil-stock free, nor does it necessarily mean giving up a mining royalty either.

Humanity might be "creating a ticking time bomb," but Gore's home in Carthage is sitting on a zinc mine. Gore receives $20,000 a year in royalties from Pasminco Zinc, which operates a zinc concession on his property. Tennessee has cited the company for adding large quantities of barium, iron and zinc to the nearby Caney Fork River.

The issue here is not simply Gore's hypocrisy; it's a question of credibility. If he genuinely believes the apocalyptic vision he has put forth and calls for radical changes in the way other people live, why hasn't he made any radical change in his life? Giving up the zinc mine or one of his homes is not asking much, given that he wants the rest of us to radically change our live styles.

The Debate is Over?

According to Albert Gore Jr. "the debate is over." Really now, since when has science stopped progressing, challenging itself, re-discovering itself? Never, because that is not the nature of true science, disciplined scientists never cease to question known beliefs or postulations. The search for truth and understanding is an ongoing process not to be dictated by a few arrogant elitists.

Real science never stands pat. Never before has every single scientist on the planet come to a consensus on any theory of this magnitude. You can always find some scientists willing to challenge just about any established theory.

The nature of science is skepticism; by challenging and re-challenging the so-called accepted positions, we all ensure sound science prevails.

Curiosity may have killed a cat or two, but it drives science to new levels of repute. It forces anyone (including failed politicians) making claims of significance to prove it and back it up with credible data. Then, one must be able to recreate the hypothesis in a controlled environment until anyone in a middle school science class can demonstrate the proof.

I know of scientists still attempting to poke holes in the theory of gravity! Good luck, but that is what makes science re-assuring. It is never satisfied. The debate may be over in Al Gore Jr.'s narrow mind, but the search for the truth is always active in the conscience of every true scientist.

Frankly, I am shocked that so many scientists are signing on to a theory that is so flimsy and nonsensical. I am aware of many scientists who are now backing down from the doomsday analogy the alarmists are spewing - with good reason. There is little rationale behind their fear mongering. However, ulterior motives will be explored in later chapters. Please keep reading…

Let me remind you that I am not denouncing Al Gore Jr. personally because I disagree with his politics. I am adamant about undoing the crime he is committing. It is frustrating to see a public figure be so hypocritical and incredulous about the consequences of his newfound religion called "Global Warming."

Instead of defending their claim, alarmists (Gore in particular) have turned critical of anyone who repudiates their claims. I look forward to standing on the other side while the global warming apocalypse goes down in flames. My salvo to Al Gore Jr. is, "If you can't stand the heat, stop talking. You are spewing CO_2!"

Another liberal trick employed by politicians is to discredit their critics -- blame them for doing what the politicians themselves are doing. If the global warming lie, I mean theory, is true, then it should hold up to the most concerted rebukes. If you declare the debate is over, then you are admitting your theory cannot be held up for scrutiny.

I invite you to continue your journey for the truth. When data is suspect, the gravity of the charge so deep, and the head simpleton's motives are in doubt, it is time to put on your thinking cap and get back to fundamental common sense reasoning.

Quest for Power and $$

What other reasons could Al Gore Jr. and his minions have to declare that the debate on global warming is over? Could it be now that enough people are snookered and aligned with their cause, so the elitists can move forward implementing their agenda of government control?

I am not a conspiracy theorist, but I am a realist. I can see in my lifetime how the government has transformed itself into a huge intrusive conglomerate. Ah, the good old days, when government did not tell you what you could eat, what you could say in public, what kind of car you could drive. Now the government even wants to regulate how much salt you can eat and whether your kids can get a toy with their meals. When will the nanny state regime stop…?

Today, more than ever, the government needs more massive taxes to feed itself (see national debt). Meanwhile politicians increase the regulations they impose on your property, investments, businesses and schools, as they will not stop until everyone falls under their control. We are becoming a nation of two classes: the takers and the taxed. Once the majority of "takers" exceed 50%, watch out! By design the "taxed" will be the minority and thus less likely to stave off more taxes, more regulations, and more government dominance.

Absolute power corrupts absolutely. The government has gotten this behemoth because the citizens of this great nation have fallen for the hook, line and *stinker* that government can solve all the problems. I am sorry to tell you, it cannot. It has consistently proven with its attempts at social engineering, progressive tax codes, and boorish oversight that it is the most inefficient, loophole riddled, unfair, and shameless machine.

Owning the power to influence individuals and businesses, the power to implement more government programs, and the power to extort more money from individuals and businesses is exactly what all egotistical elitist dream about. Do you really believe that the politicians who levy more taxes and enforce porous regulations have the purest of intentions? If you buy that nonsense, you will see just how subservient all Americans, including businesses and individuals, become to these tyrannical *uncivil* servants.

Money in politics has risen to astronomic levels because the stakes have risen as well. Even at the local level, I see this all the time. A business wants to build on an empty lot."Oh no! You will ruin the pristine nature of that mud hole over there," decries the zoning board. But amazingly, after a million dollar donation to enhance the nearby roadways is added to the plan, the zoning board welcomes the business with open arms. I call it legalized extortion: technically legal, but ethically depraved.

Let's focus on the global warming agenda of more control over private business and free individuals. What do Al Gore Jr. and other likeminded political figures gain by increasing your taxes? What do they attain by having more and more regulations over individuals and businesses? Obviously, this entails the acquisition of power; power to create and impose the perverse ideals of liberalism onto unsuspecting people. Then the 20% of liberals, claiming a progressive mandate, attempt to squelch those in favor of the U.S. Constitution (i.e., liberty, commerce, self determination).

Now that Al Gore Jr. is highly unlikely to attain his ultimate goal as commander and chief, it is evident he is decided to reap the rewards of our capitalist system, except he is rigging the game in his favor there too. While altruistically promoting a "green" agenda, he has his money on the table – big time! Did you know going green is now big business? Many corporations are anticipating huge profits from mandated government initiatives and Gore is invested in many of them. Why is anyone surprised that Al is cashing in on the very fear-mongering he personified.

You need proof?

A March 6, 2009 Bloomberg News story reports:

Former U.S. Vice President Al Gore left the White House seven years ago with less than $2 million in assets, including a Virginia home and the family farm in Tennessee. Now he's making enough to put $35 million in hedge funds and other private partnerships.

The Bloomberg piece continues:

He and Tipper Gore released tax returns for 1998 showing they
earned $224,376 that year, less than half the income of President
Hillary and Bill Clinton, news reports at the time said.
Now Gore charges a $175,000 speaking fee and has a reported
net worth "well in excess" of $500 million.

Mr. Gore's firm, Kleiner Perkins Caufield & Byers, one of
Silicon Valley's top venture capital providers, is looking for $75
million to expand its partnerships with utilities seeking to install
millions of so-called smart meters in homes and businesses.
Coincidently, the Energy Department announced $3.4 billion in
smart grid grants, the New York Times reports. Of the total, more
than $560 million went to utilities with which Silver Spring has
contracts. The move means that venture capital company Kleiner
Perkins and its partners, including Mr. Gore, could recoup their
investment many times over in coming years. When Mr. Gore
says he is "going green", does he mean he is pursuing ecological-
friendliness or economical-friendliness? Regardless of his
motives, he is cashing in – big time!

*"Do you think there is something wrong with being active in
business in this country? I am proud of it. I am proud of it."*
Al Gore, Jr. 2009

Global Warming Myths

Myth #1 (straight from Al Gore's web site)

Greenland is melting (so say the shrieking climatologists reminiscent of the wicked witch of the west). On Al Gore's website (don't bother surfing, it may melt your eyes), he claims that the flow of water from Greenland's glaciers has more than doubled in the past decade. Greenland does, in fact, have a large amount of glaciers. In a shortsighted world, this might be of some alarm.

However, nature runs in cycles. Thus, in order for Greenland to have glaciers in the first place, at some point it must have been twice as cold to accumulate such a net gain of ice. Was this the impetus of the global cooling scare of 1926? It's like saying the snow melting in the spring is a disastrous event in and by itself. Until you consider the accumulation of snow and ice each winter, it may seem that way to a shortsighted person.

The reality is that Greenland would take thousands of years to melt. Even if the edges are reportedly melting, the middle ice of Greenland is in fact growing. Consider the fact that the continents are still moving, the oceans may rise an inch or two over the next century, and it was warmer in the 1930's and 40's than it is today. So don't let the global alarmists get you down. Their predictions change with the wind and now it seems with the temperature.

I would be remiss if I did not state for the record how clever it was to name a virtual island of snow and ice "Greenland" and have the forethought to name the cozy, hot springs spa island "Iceland." I would like to discuss that with their original travel agent...

Myth #2 (Also, straight from Al Gore's Web site)

Malaria has spread to higher altitudes. I am not an expert on insects, and neither is Al Gore. But common sense tells me that perhaps the insects have adapted to fly into higher elevations. Insects have been around for a very long time, and it always amazes me how they manage to creep into my basement. Is malaria restricted to one kind of insect now? Perhaps a hardier breed is now carrying the disease. How about the fact more people live in higher elevations now? Doesn't that increase the odds of more malaria outbreaks at higher elevations (logically speaking, of course)?

But, I guess to Al Gore Jr. it is an undeniable truth that the .5 degree increase in temperature has caused these particular disease-riddled mosquitoes to fly higher than an eagle (warm air does rise), thus explaining all the outbreaks on the mountain tops. I am all for stomping out malaria; I would just rather give them all a vaccine rather than shut down what is left of our manufacturing industry here in America. Just call me a crazy conservative for thinking that way…

Myth #3 (Yep, right from Al Gore's Web site)

279 species of plants and animals are moving closer to the poles in response to global warming. So what? Plants and animals spread all the time, especially the migrating ones. Perhaps these 279 plants and animals are moving because they have instincts and survival mechanisms to encourage them to expand. All localized life is in jeopardy if it continues to remain in a small area; that is why all plant and animal life tries to expand -- it is called thriving! Notice the species called *silver haired senior citizens*; thousands of them flock to Florida every winter. I believe they are attracted to the prey know as the *early bird special.*

Plants and animals throughout history have traveled across oceans and continents, as means of survival. This planet would be void of life if they didn't branch out, adapt to changing environments and evolve into stronger species; this is Biology 101! Every ecosystem is continuously changing -- from day to night, from summer to winter, and from generation to generation. Plants and animals will continue to struggle, adapt and evolve, much to the chagrin of liberals who demand the Earth stand still long enough for the census bureau to count every living thing.

Myth #4 (You guessed it, straight from Al Gore's Web site)

Deaths from global warming will double in twenty-five years; that prediction better start turning around soon because, in reality, deaths from natural disasters have gone down over the last decade. That may seem hard to believe with all the sensationalism produced by the media, but the fact remains that fewer people have died as a result of natural disasters [1]. Credit this fact to better technology and communications from early warning systems. How many people died as a direct result of hurricane Katrina? 1723 people died which is not insignificant. But when the news reports:

Tuesday, August 22, 2006
Update: The unofficial Katrina death toll has risen from 1,723 to 4,098 as of March 13, 2007. See my March 30, 2007 post, **Katrina Death Toll Passes 4,000** for details. Then we hear **Katrina Death Toll Plummets to**

It makes you wonder what the press wants you to believe. Their tactic is to use sensationalism first, and the retractions can come later (usually smaller print). Why not launch a global study into how many news stories have been forced to retract claims because of shoddy journalism. Now that would be an earth-shattering report.

Also, the fallacy of the increasing death trend is also true for ice and snow related accidents/mortalities, as they have dropped significantly, too [2]. It is the hope of every winter white out zone, that if the warming trend does continue, real life savings will continue to increase as well. With less ice and snow covering the roads, less energy (mainly diesel guzzling snow plows) and metal corroding salt will be expended to combat these cold weather conditions. As long as a .5-degree temperature change does not claim a half million heat stroke victims, we will certainly have a net gain.

Myth #5 (Where did I get this? Oh yeah, right, from Al Gore's Web site)

Global sea levels could rise 20 feet if Greenland and the Antarctic both melt. Notice the words "could" and "if" and "both". Change them to "will", "when" and "do", and then I will jump on the *ban* wagon. Until then, it is misinformation designed to sell you a bill of goods or actually, a bill of more government regulations and taxation…

The "coulda," "woulda," "shoulda" lingo is a classic grammar trick targeted at feeble minds which often glance over the grounded prepositions, but panic over the lofty adjectives. Dangling participles ain't gonna cut it with me!

Here is my prediction: *"Global alarmists might gain some credibility if they told us why they really want to scare everyone"*.

Myth #6

Greenhouse gases trap the Earth's heat. Yes, they can, but they can also reflect more sunlight, thus cooling the Earth. Try this experiment for yourselves: Fill one container with regular air and another with the greenhouse gases of your choice. Put them both in the sunlight and measure their temperatures. My prediction is that you will find no measurable difference unless you are a liberal scientist looking to get more funding to further your *objective* research.

Have you all gotten tired of the politician's line, "We must do it for our children?" Well, here we go again, but this time it is, "We must raise taxes and increase government controls for the good of the environment; oh yes, and the children too…"

Our government spent over $25 billion since 1990 on global warming research, According to the GAO. You tell me it is not big business.

"Common sense ain't common."
Will Rogers

Chapter 5   The Lie

Let's take a quick look at the history of the environmental movement, because I see a distinct pattern emerging. According to environmentalists, we are always doing something destructive... first, in the 70's it was global cooling, and then came pollution, and the mean old loggers destroying the forests. After the acid rain theory was debunked, they moved onto man-made chemicals depleting the ozone layer, and now this week its global warming.

We survived the pollution, the forest devastations, refilled the ozone hole and miraculously turned around the global cooling so well, it swung us into global heating! This one has taken root for the whackos in part because it is not quantifiable. Also, it will be another hundred years before my great grandchildren can say, "Silly us, one to two degrees warmer actually benefited the planet."

The Holy Ozone

I truly believe this global warming movement stems from the success that the ozone hole alarmists generated back in the 80's. At least in my lifetime, no environmental movement has had so much success in its ability to scare the public and initiate sweeping regulations on businesses. The ozone depletion fear mongering was a complete victory for the environmentalist whackos.

Earth is not constant, perhaps because it is spinning. Seriously, it never has been static and never will be. I am sure it will continue to change whether we are exhaling CO_2 or not.

Everyone who studies the atmosphere knows the ozone at the poles increases and decreases yearly like just about everything else in nature. However, those early environmentalists scarred enough unsuspecting voters and empowered enough Democrat politicians to enact legislation that forever changed the auto and refrigeration industry.

From every American product, fluorocarbons were banned like Typhoid Mary at a cocktail party. The ozone scare was a costly escapade to those related industries. However, it will pale in terms of cost to this new environmental scare tactic that is now gaining steam and has set its sights on devastating what's left of the American manufacturing sector. No doubt, the ozone hole *win* has emboldened the new environmental alarmists to reach for the sky, literally.

The Real Inconvenient Truth is Reason

For anyone who lives on the liberal "dark" side, they must abandon rationale or at least develop tunnel vision. To get snookered into another dreary theory is to abandon common sense. Fool me once, shame on me; fool me twice, shame on me even more! It is time to wake up to the socialist movement called global warming. Let this be the alarm bell ringing. Once you have finished reading all chapters, sleep tight and feel free to hit the snooze button in the morning - Global Warming is a hoax!

I, for one, want to enlighten people, but the real success would be encouraging people to enlighten themselves. We all have the facilities to reason. We all have experiences to draw from. Furthermore, we all have access to trustworthy people with the capacity to reason, coupled with the ability to communicate. Do your homework. Know the source and understand their motivations.

We all have the ability to see the glass half full or half empty. Take hope, derived from reason, and let that fuel your optimism that we shall overcome all obstacles. As a conservative, that is our hallmark. We conservatives believe in the power of individualism and entrepreneurship. Given enough freedom, America has proven what great thinkers have known - allow people to reap the bounties of their labor and look out, because most people will seize the opportunity and do great things. Everyone has the inalienable right to be free.

Our government has not made this country the most successful of all time. Arguably it was and still is individual's ingenuity, determination and perseverance. The most successful eras have been when government has gotten out of the way of free enterprise. When risk is rewarded, when labor is compensated and rights of ownership protected alongside the moral rule of law in conjunction with a touch of humility (thankfulness), good things will happen...

"Common sense is the best sense I know of. "
Lord Chesterfield

Liberal Dinosaurs

Good thing for us the dinosaurs didn't have a liberal controlled government. We can be certain if they did that the liberals would have prevented their extinction, right? Of course, because the liberals upon learning of the impending meteor shower would have leapt into action: their first response would be to raise taxes. This would be consistent with today's liberals.

Secondly, they would have taken a poll to see what the popular course of action should be. Remember there were no term limits back then either (except for well placed meteors).

Lastly, the liberals would have commissioned a blue ribbon panel of experts (i.e., campaign contributors) to determine where the meteors would land. Then, depending on the voter demographics (for example, if the target area was heavily liberal), propose moving everyone to higher ground. If, however, the target areas were conservative, then it would be a local issue and the people would have responsibility to save themselves. Either way, the proposals correlate nicely with that of today's liberals.

Seriously, it is imperative to understand the liberal mindset (even if you are a liberal). Their method is to create fear, because without fear, people tend to think and act rationally. Put fear in their hearts and tug at their emotions, or worse, threaten their very survival and look out! You can convince every malcontent, day laboring, and dope-smoking protester to rally behind your destructive policies! Policies like:

- Higher taxes to pay for more government programs (a bureaucrat's dream)
- More regulations to squeeze campaign dollars from businesses (legalized extortion)

- Less individual freedoms (keeps people dependent)
- More power to egomaniacs in office (more prestigious awards, too)

Conservatives want to expose these charlatans, while liberals want to shower them with phony accolades and prizes. The mindset is that if you support the leftist cause, you must care. You automatically become compassionate and wise to the plight of the poor and downtrodden. On the other hand, conservatism is reserved for the wealthy well-read types who selfishly plunder and hoard their fortunes.

The democrats are the party of "good intentions." Why do you think the liberals promote dependency on the government instead of giving incentives for healthy, capable people to work? Who votes in droves for liberal Democrats that support these entitlements? Is it the politicians or the "me first" voters fault for flagging the system? The answer is both!

It is obvious liberals desire to increase their voting base at the expense of the working middle and upper classes. I never understood how anyone who earns their own keep could justify voting liberal, unless they did not see the utter destructiveness of government dependency and over regulation. Otherwise, a person would have to be sadistic or masochistic to perpetuate big government.

Liberals can create a victim out of any group of people: women, minorities, and even the Islamic Jihadist. (Why can't airport security single out middle-eastern men between the ages of 17-45?) In order to be fair, liberal rule-makers prefer the TSA strip-search grannies and infants. How Idiotic! So in order not to offend the terrorists, passengers keep taking off your shoes and leave your shampoo at home.

Trick or Treat

How do we know that the global warming theory is a liberal trick or treat? I think it is more of a trick and less of a treat. So let's look at the benefits going to the winners of this inconvenient truth.

Global Tax: Liberals get more money, control over individuals, and become one step closer to a new one-world order. Individuals have less money to invest, give away, or otherwise spend.

Government Regulation: Liberals gain more control over businesses and easier access to campaign contributions. Individuals have less freedom to grow a business weighed down by more paperwork.

Responsibility: Liberals mandate government to assume risks and responsibilities for the individuals, creating more dependency on a bigger government. Individuals will have less responsibility to provide for their families and themselves with expectations for more and more government aid.

Traditions: Liberals replace American traditions and values with a politically correct code of fairness (non judgmental). Individuals forget how and why we became such a successfully wealthy nation and soon lose our identity to a powerful central government.

What better way to build on an all-encompassing, restrictive, omnipotent government than manufacturing a doomsday scenario that justifies higher taxes, more regulation and bigger central government? Notice Al Gore, Jr. and the like never state any faith in individuals or the free market to solve problems. Nope. Only big government can save the day. Very shortsighted, and for good reason, the Elites don't prosper from less government and individual successes, except through higher tax revenues.

The world according to liberals relies solely on the virtues and efficiency of big bureaucracies to try and solve the planet-in-peril scenarios. The key word here is, try, because the government really never solves any problems. Can you name a social issue the government has completely solved? Even if you could conjure up a program, ask, "What was the overall cost?" "What were the unintended consequences?" When it comes to government programs, usually the costs do outweigh any benefits.

The mark of a successful program is that it targets a problem, has the correct funding, and executes a detailed plan. Then upon solving the problem, it becomes disbanded. Now I know that you won't be able to conceive a government program that fits this bill...

We all know liberal government leaders like to throw money (or IOUs) and regulations at problems. Unfortunately, the problems only get worse (e.g., consider the war on poverty). Campaign pledges (i.e., buying votes and favors) and implementing unwarranted policies (i.e., paying off donators) leads to huge bureaucracies so entrenched that ending the program would create a whole new catastrophic event in and of itself. Now most government programs are "too big to fail".

I would like to tell other fellow citizens about our New York state leaders and the NY Thruway Authority. In case you were unaware, we New Yorkers are one a few states employing a toll system across our Empire state. A 20-year bond was secured in the 1980's to pave the thruway along with the government's promise to remove the tolls upon the expiration of the bonds. Nice selling point, but too bad few can remember 20 years of state legislation history.

If you have visited New York, you still see the tolls in place, fully stocked with union toll takers. The current government could not keep its promise, as that would have displaced too many toll workers and without the toll money coming in they couldn't pay the bloated Thruway Authority's overhead. So, to add insult to injury, the all-powerful and unapologetic authority, requested and got a toll-rate increase instead of keeping its promise and moving to the real world of free enterprise.

The Republican run government spent in excess of one and a half billion tax dollars per year on climate research [3]. Let's look ahead, shall we? What would you expect to happen as big-government liberals are at the helm? Try five billion under Obama's first budget. Even more money will sail out of your pockets and into this fiasco until people vote STOP! The *pirates of the DC* will need to hire and train an entirely new fleet of recruits to terrorize, I mean regulate, everyone emitting CO_2 (that would be everyone still breathing.) If we could only turn this CO_2 Gestapo against Al Qaeda, we'd scare the hell out of the terrorists!

Here are some predictions that came true about what liberal politicians have proposed once becoming elected:

In order to save the planet, we need to:

- Raise taxes to hire 16,000 CO_2 police (I mean IRS agents)
- Ban all SUVs, except those of Government officials and Hollywood movie stars who contribute to the DNC
- Increase regulations on what's left of our manufacturing base until all polluters are safely overseas
- Sign on to a global treaty that only punishes the wealthy nations and exempts the worst polluters (holding our debt does have its privileges)

- Grant more money for research into why the sea kelp is 15% less tasty than last year and now according to the FDA, Americans eat too much salt

Once more taxes are imposed, tougher regulations are in place, and free markets are controlled by the central government, do you think for a minute the problem of excessive greenhouse gases will get solved? Do you think that then the leaders will announce success and begin to cut the programs? The only thing that will change is that taxpayers will be poorer, the market will be smaller and government will be larger, making politicians even that more dangerous.

Can you imagine any government advocate announcing that the global warming issue has been resolved, ever! If you believe in unlikelihoods, then you may also believe the government will begin repealing your tax increases, stem the unnecessary regulations, and reduce the size of government. Sorry, this has only happened once in our 200+ history (thanks **"Ronald Regan"**) and appears to be increasingly unlikely to reoccur anytime soon, unless the TEA Party gets its way...

Results Matter!

Results do matter! If you don't believe me, see the results of the war on poverty. To a liberal, it only matters that your intentions were good. We live in an era where results do matter -- where consequences from actions matter. The stakes are getting higher as the government has seen fit to grow into the giant bureaucratic behemoth it is.

The governmental accounting department no longer rounds dollars to the nearest thousand. No, now it is to the nearest hundred thousand dollars; check the latest budget synopsis. Most people cannot fathom three trillion dollars (approximate budget for 2009). Knowing a small congressional committee is in charge of billions of dollars, are you scared? What is more likely to bring down this great nation - a minor temperature fluctuation or a corrupt and morally bankrupt government? Why do you think liberals are intent to take away your guns?

What is more likely to cause a global meltdown - global warming or a rogue government? Look at North Korea, Syria, Iran and China -- it is not the people we fear, it is the lunatic leaders with tyrannical authority. Look at the hurt they are inflicting on the US economy. With a few countries holding our oil supply hostage, while a few others threaten to call in our massive debt, you tell me what is scarier - economic chaos or a piece of ice melting in Greenland?

The reason I bring this into discussion is not to scare you (enough of that in the movies), but rather to demonstrate the real problems and challenges we must face together. These issues have real consequences, but are being deflected by the environmentalist whackos (as Rush likes to call them). Sadly, most liberals can't deal with these global issues until they are so completely out of hand, at which point they claim to have solutions for the very problems they created by inaction. How can we let these liberals be in charge of world affairs tomorrow when they cannot even see the real dangers staring them in the face today?

Notice how the strategy works for liberals - first, create an atmosphere of resentful class warfare. Next, promote the end-of-the-world scenario so you can rise to power as the global saviors at the expense of the few successful Americans who already fund a majority of the tax base. Then, continue instituting liberal reconstruction, or as I call it, "deconstruction" of the basic American tenants and customs. Finally, diminish the liberal's greatest foe, organized religion, thru political correctness. Once the Oligarchy is achieved, cement the bricks of socialism until it cannot be repealed.

The bottom line is fear is the tactic that government elitists use to increase power and decrease your freedoms. Once you lose them, you rarely ever get them back. Short of a coup, when is the last time the government ever reinstated a freedom it took away previously?

Propa-scam-da

Before allowing it to be shown in schools Britain-wide, the British government has officially tried Al Gore's global warming *propa-scam-da* film, "An Inconvenient Truth," in court and has reached the following conclusions:

_____

The court found that the film was misleading in eleven respects and that the "Guidance Notes" drafted by the Education Secretary's advisors served only to exacerbate the political propaganda in the film." [4]

In order for the film to be shown, the government must first amend their Guidance Notes to Teachers to make clear that:

The film is a political work and promotes only one side of the argument.

If teachers present the film without making this plain they may be in breach of section 406 of the Education Act 1996 and guilty of political indoctrination.

Eleven inaccuracies have to be specifically drawn to the attention of school children.

The inaccuracies are:

1. The film claims that melting snows on Mount Kilimanjaro evidence global warming. The government's expert was forced to concede that **this is not correct**.
2. The film suggests that evidence from ice cores proves that rising CO_2 caused temperature increases over a period of 650,000 years. The Court found that the film was misleading: **over that period the rises in CO_2 lagged behind the temperature rises by 800-2000 years**.
3. The film uses emotive images of Hurricane Katrina and suggests that this has been caused by global warming. The government's expert had to accept that it was **"not possible"** to attribute one-off events to global warming.
4. The film shows the drying up of Lake Chad and claims that this was caused by global warming. The government's expert had to accept that **this was not the case**.
5. The film claims that a study showed that polar bears had drowned due to disappearing arctic ice. It turned out that Al Gore had misread the study: **in fact four polar bears drowned and this was because of a particularly violent storm.**
6. The film threatens that global warming could stop the Gulf Stream, throwing Europe into an ice age: the claimant's evidence was that this was a **scientific impossibility.**
7. The film blames global warming for species losses including coral reef bleaching. The government **could not find any evidence to support this claim.**

8. The film suggests that the Greenland ice covering could melt causing sea levels to rise dangerously. **The evidence is that Greenland will not melt for millennia.**
9. The film suggests that the Antarctic ice covering is melting; **the evidence was that *it is in fact increasing.*** [4]
10. The film suggests that sea levels could rise by 7 meters causing the displacement of millions of people. In fact **the evidence is that sea levels are expected to rise by about 40cm over the next hundred years** and that there is no such threat of massive migration.
11. The film claims that rising sea levels has caused the evacuation of certain Pacific islands to New Zealand. **The government is unable to substantiate this** and the court observed that **this appears to be a false claim.** [5]

Basically, I think all it does is make Albert Gore Jr. even richer from his carbon trading company. This is especially poignant since Gore himself doesn't cut his own CO_2 levels, preferring instead to offset them with carbon credits… purchased with his money… which he earns from carbon credits. Wait until Goldman Sachs gets a hold of this derivative…

Imagine if you were a has-been politician with no legitimate legacy? Imagine an egomaniacal elitist who needs the spotlight, but cannot get back into the political arena. What better way to lift your sagging spirits than to lead a worldwide effort to impose your sanctimonious views on mankind, while empowering your fellow socialists to begin the process of increased taxes and over regulation? That is the Democrats' solution to every problem, right? Have you ever heard a liberal clamoring for fewer government regulations and lower taxes for all taxpayers? They are the new elitists who know what is good for us poor slobs, and we should be thankful they are there to save us from our nice houses, our big cars, and now apparently ourselves.

"Remember, random searches only work randomly."
Unknown

Chapter 6   The New Old Media

The mainstream media (e.g., network news, magazines and newspapers) has changed and not for the better. That is why the once powerful media is now the "Old" media; just as vinyl records gave way to streaming music, the dominating newsroom is giving away to cable and internet free-thinkers. Listeners/readers are no longer fooled by the sometimes subtle, other times blatant, biases of the "Old" media.

Fortunately, fewer and fewer people rely on the mainstream media for their version of the news (see plummeting ratings). Now that they no longer have a monopoly due to the success of cable, talk radio and the Internet, it is evident they have kicked into survival mode; unfortunately their aging subscribers are not following suit.

So, while the "New" media continues to gain market share via cable, Internet and radio, to a more "in-tune" audience, the "Old" mainstream media continues to desperately flounder. They flail about like a fish out of water, seeking the most extreme stories and exaggerating their minority views, while speculating and editorializing instead of reporting the facts. They have brought storyline sensationalism to an art form. As the new generation of thinkers migrates to the new digital mediums, the old fashion analog format will continue to fade into static.

"Common sense is very uncommon."
Horace Greeley

Bias Media or Media Bias?

Television news is big business (just ask how much the anchors get paid), and they will do just about anything for ratings -- I mean anything. Look at the cliff dive Dan Rather took over the phony National Guard records he produced, thinking it would bring down George Bush. Sorry, Dan, you need to investigate your sources before blabbing untruths to your audience. Enjoy your retirement; I know I am.

Isn't it possible that the highly competitive media are similar to the popular weather channel programs airing 24/7, which harps on every little weather anomaly everywhere in the world trying to keep watchers interested? I can't watch these weather shows for more than a minute or two. But I have noticed the reporters get real excited announcing that a potential storm is brewing, or calculating the damage of a particular impending weather calamity. What would you say or do if that were your job and you really wanted to keep it?

Just one more good thing I want to bring to light: GE once owned NBC. NBC went "green". GE sells $700,000 dollar turbines. Would you be surprised to learn that GE benefits greatly by the increased sales of its alternative energy products? I don't blame GE for trying; it's like having a commercial ad running 24/7...

The mainstream media and weather forecasters in particular are guilty of a subtle psychological trick that they play on unsuspecting viewers: selective reinforcement. This is the continual highlighting of the particular weather phenomenon that they want to convince you its predominance. For example, if the weather forecasters want you to think it is cloudier than, say, twenty years ago, they will continuously point out those occurrences when it is actually cloudy. It may have been sunny all week long, but as soon as it becomes cloudy, it is over emphasized and promoted as supporting evidence for their claims.

Contrarily, when it is not cloudy, the entire subject of excessive cloudiness is ignored. I can prove this is *in play* with global warming... When is the last time global warming was mentioned during a cold spurt? Contrarily, wasn't it the headline story whenever it became unseasonably warm? Of course it was! That is how selective reinforcement convinces you the worse is happening. The worse is happening, the worse is happening, it's getting worser and worser... See I just convinced you!

The Weather Channels

The increasing sensationalism of the weather by competing networks and cable shows magnifies the ordinary weather events to the extreme. I could also convince you that the political bias of the news could accentuate weather events like, say, Katrina! Never before has a weather phenomenon caused so much political overtones by weather forecasters / reporters. It did not take long for a horrific weather calamity to become a political spin machine by the supposedly impartial mainstream media. Compare Bush's Katrina to Obama's Oil Spill; the contrast by mainstream media could not be clearer...

How accurate is the climate data spewed by the forecasters? The data used to support global warming, the basis that reputably seals our fate, is <u>not</u> without question. How reliable is the weather data from a hundred years ago? Even if the U.S. data is reasonably sound, what about other parts of the world?

How good were the weather monitors in South America a century ago? Or Africa and China for that matter? Don't tell me they are basing their global meltdown theory on just U.S. and some Nordic weather data!

The weathermen back then, certainly more respected than today's prognosticators, did not have the technology we have today, nor did they keep meticulous data like we do today. Nevertheless, "the consensus" experts calculate all the data, past and recent, the same. I am sorry; I 'm not planning my funeral based on data collected by horse and buggy novices. Nothing against novices, they just do not garner the same high standards required when it comes to life-altering theories and speculations that threaten my liberties.

Here is an interesting fact to consider when analyzing global warming weather data (like most people have the time for such amusement): when the Soviet Union collapsed at the end The Cold War, many of the weather stations across Siberia stopped collecting data. Coincidently, the subsequent years of global weather data began to show an increase in temperature. Amazing how these things work out for the alarmists.

You know that when the experts announce polling or statistical data, they always specify a margin of error, right? Well, what is the margin of error for our irrefutable weather data? With so much guesswork, variables and ever-changing "constants," there must be some margin of error! Well, I have yet to hear of any pronunciations of what the real margin for error really is. There are two likely reasons for this: they're so arrogant they believe they can foretell the weather accurately fifty years from now or the margin of error is equal to or greater than their predicted increase. So, **if** their one or two degree increase over the next hundred years was marked with a margin of error of, say, two degrees, who would panic? No one in their right mind, because the prediction becomes statistically meaningless! Don't hand over your car keys just yet?

First, I would like to stipulate that many weather forecasters are in collusion with the global alarmist either by ignorance or for profit. Profit? How can forecasters profit by global warming? Easy, the ratings increase with the sensationalizing of the weather. Do people tune into the weather channel when it is 72 degrees and sunny or when it is near freezing and a nearby storm front is moving in? If their bias is not motivated by profit, then, unfortunately for them and us, it is pure ignorance that is at work here. Perhaps if we hack into some of their Emails we would find the truth of the matter.

Your Guess is as Good as Mine

Ever get sick and tired of hearing the weather person pronounce the week's forecast as partly cloudy and partly sunny? This is kind of a cop out if you ask me. It may rain or it may not. Seems like the only thing they are sure of is that the Sun will rise at dawn (something AM) and set in the evening (something PM). They are extremely accurate at predicting that, have you noticed?

Getting back to cloud cover; have you ever been outside and felt the heat from the sunshine? Then, all of a sudden, the clouds interrupt the light rays and you immediately feel cooler. Makes sense that cloud cover would affect the global temperature, right? Absolutely!

The problem is cloud cover is too unpredictable even with all the technology we have today. So how does a global warming scientist calculate future cloud cover's impact on the rise or decline of global temperatures? Sorry, they cannot even predict cloud cover next week, next year or the next decade any more than they can predict the direction of the stock market!

Why does it seem the weather is getting more chaotic and more destructive than ever before? There is a simple explanation. Again, stepping back and analyzing the perceived dilemma in a calm, rational way unravels the mystery of the catastrophic climate conditions (i.e., CCC). First you need to understand that there are more people in more places with more property than ever before. Also, the dollars at risk are enormous as our affluent society builds more waterfront houses the size of Al Gore's homes.

Three hundred years ago, when a category five hurricane struck the everglades of Florida, not too many heard about it. Nowadays, there are so many people in central Florida that any whirlwind sighting is seen as a potentially devastating threat to that area. Add a camera to everyone's cell phone, instant messaging of every little event and bam – news of every weather related incident is available to the public.

Also, due to a large population of older people with a peculiar interest in weather-related news, more and more coverage is dedicated to every single storm or even tropical depression. (What is depressing is the incessant hyping of every gust of wind!)

All this new technology leads to extreme competitiveness of reporting, which can lead to exaggerations. You think the health food industry is the only one who exaggerates? Media sensationalism has increased exponentially in correlation with the increase in cable and Internet news. Mainstream media competition is fierce, and the stakes are high. The pressure to justify their million dollar salaries must be enormous...

My claims here about weather journalists may at times be subjective, but what does your common sense tell you?

Droughts and Wildfires, Oh My!

Scarier than the wicked witch of the west, the global alarmists warn that we will experience more droughts and endure more wildfires around the globe in the next several years. I actually agree with this claim. Surprised? Don't be. Droughts may be increasing, but not because of a minor temperature increases (purportedly caused by CO_2 in the air). Rather, we are using and diverting more and more fresh water. The planet's population is growing and so is the need for fresh water. We have been depleting underground fresh water reserves for years, but this has nothing to do with the proclaimed climate changes.

Droughts and rationing have been occurring and will continue to occur regardless of the CO_2 levels. Every part of the United States keeps statistics on rainfall, and I guarantee that every year they vary. Just as usage and demand vary, fresh water supplies will change, too. Again, we live in a non-static world. It rains more some years than others, and this has been going on long before man invented the aqueduct.

Fortunately for us, the planet has more water (about 70%) than land. Eventually, we will someday be forced to desalinate the seawater to supplement our water needs. This is a growing vital industry. Fortunately, our increased water consumption, of course, will offset the melting glaciers and save our coastal regions from imminent flooding. Maybe some enterprising person will bottle and sell the melted glacier water, with the slogan "Save the world – drink glacier water!" So, drink up everyone!

As for wildfires, they too have been around for a long, long time. It is the way nature recycles itself and nothing to do with CO_2 emissions. The ferocity of today's fires does have to do with our land management policies. In particular, our unwillingness to permit cutting safety zones and not allowing to clear underbrush has caused fires to be more destructive than necessary.

Next time there is a wildfire, pay attention to the coverage by the mainstream media. Out of the entire horde of investigative journalists that descend on these annual occurrences, rarely does one mention the root cause: environmental whackos who prevent clearing of the undergrowth or buffer cutting the trees to help stop wild fires from spreading in the first place!

The increase in dead brush acts like kindling in the arid regions and increases the intensity and speed in which wildfires devastate the land. Furthermore, there are many ways to create safe zones by carefully clearing areas of high risk. But, "No" say the tree huggers, as ironically they allow thousands of plants and animals to suffer the ravages of fire and only curse when millions of dollars in property damages occur under their watch.

It is time to push these idiots off the cliff and restore some intelligence back onto our land management teams. My new slogan is "Save our forests -- plant some intelligence and cut down the weeds of environmental zealots." It may be too long for a bumper sticker, but just right for this book.

Global alarmists want to convict industrialized nations for the tsunamis, wildfires and every other natural disaster on the globe; the problem is that the facts get in the way. This crazed mob of malcontents wants you to believe that we are to blame, so that the endless inferences and allegations towards blaming global warming on man are constantly cited until absorbed into the mass psyche. The repetition of these assertions is consistently spewed out as fact is a vain attempt to support their madcap claims.

It does not take a geologist or even a gemologist to explain that earthquakes under a large body of water cause tsunamis. Furthermore, earthquakes have been happening for a long time, well before acid rain and chlorofluorocarbons. Additionally, lightning is usually the cause of forest fires, not the overabundance of greenhouse gases. Note: most greenhouse gases are not combustible.

Debunking Climate Change

Al Gore, Jr. cannot lose when arguing that the climate is changing; he can't because the planet's climate is always changing. Historically speaking, it has always changed, and there is no evidence to suggest that it won't continue to change in the future. At our current level of climatology, we cannot know for certain what the exact causes are for every weather phenomenon at any given moment on Earth. That reality is hard to swallow, especially by elitists who do know everything and scientists too egotistical to admit they don't know everything.

Most scientists do understand our limits, but a few others have convinced themselves they have enough data to forge conclusions that are otherwise just theories. Conclusions of this nature come from long arduous studies and analysis that would otherwise subject doubt into their conclusions. My conclusion is that we are far from making any concrete conclusions.

With the finite amount of data and our limited capacity to recreate in a lab the countless variables involved, no scientist in his right mind can say for certain that man made emissions will cause a mega disaster in the next 50-100 years. At first I thought Al Gore Jr. was out of his mind, but after seeing he was able to drop 35 million on Capricorn Investment Group, he is in a green state of mind.

Weather is cyclical on a global scale; which means local forces are not the only ones at work when it comes to predicting the weather. The fact is there are too many variables to calculate accurately the future weather. It is like calculus on steroids. We all remember how hard calculus was right?

Consider this piece of logic: Increased populations have increased the likelihood of death and damage in weather sensitive areas. High-risk areas like coastlines and the border areas between frequent cold and warm fronts seem to attract homeowners. Basically, nowhere is immune, but some places are more prone to cyclical weather anomalies than others. When a typical storm hits a populated coastal area, the damages are surely magnified, as opposed to a hundred years ago, when no one lived there at all. You see my drift. With the expansion of population comes more risk. Right or wrong, good or bad that is the reason weather damages have increased.

You cannot leave out the impact the 24/7 media has had on creating a sense that the weather is somehow inflated to levels never seen before. Just the hint of a storm brewing sends the forecasters into a flurry. Next time a tropical depression is upgraded to a tropical storm, note the meteorologists' on-air panting and excitement. I guess objectivity is now subjective…

Notice how the focus of the media does not shift to individual heroics or perseverance after a terrible storm. Nope, the focus is squarely on the government's reaction. Did the government act quickly enough? Did they do enough? Was the bottled water too warm? Were the Meals on Wheels too cold? This is the mainstream media's focus.

What did people do before FEMA? Has anyone ever survived a natural disaster before the National Guard was implemented?

I love San Diego, having visited there a couple of times. In fact, I would live there if I did not have so many roots here in the Nordic East. In 2007, I was impressed with the people of San Diego on how they handled their natural disaster of raging fires. The differences in action between the peoples of San Diego and the peoples of New Orleans were startling. If you followed both disasters, you will know what I mean when I say that I was proud to know there are Americans not waiting around for the government to save them. And I was even prouder when they did not bitch and complain when the government assistance did arrive.

The only commonality between the two events was the media's coverage. It was evident the media had to sift through hundreds of victims at **QUALCOMM** stadium until they found one old crotchety guy with a complaint. His complaint was that, when the government told him to evacuate his house, no government employee was waiting out front to drive him to the stadium. He actually had to ask a neighbor to bring him to the shelter. Imagine that -- a neighbor helping another neighbor. That is supposed to be the new role of the government, right?

Typical 6 o'clock old media -- they cannot report that thirty thousand people were well fed, sheltered and otherwise attended to. No, the *on the scene journalists* would not stop investigating until they found at least one old man with a complaint. I guess good news is not news anymore. No wonder their ratings are in a freefall.

What would your grandparents think of the new global warming alarmists? My grandparents would have told these bed wetting, communist sympathizing malcontents to get a real job! However, I must note that my grandmother would never say anything bad about anyone; she would have baked them a homemade apple pie and told them to, "Stop worrying so much." My grandfather on the other hand, would have given them all chores to do, espousing, "Hard work will make a man out of you."

If common sense won't sway you away from the alarmists, then just read a few Emails from the science community at East Anglia's Climate Research Unit **admitting climate data was hidden, others forged to make the research fit the desired results.**

The truth will be told as it is hard for some liberals to keep secrets. The more people become aware of the global hoax, the more we will learn about the cover up now taking place in certain climate research groups. As my friend Mike would say, "Follow the money."

"It is a thousand times better to have common sense without education than to have education without common sense."
Robert Green Ingersoll

Chapter 7   Liberal Leftist Alarmists

"Liberal," "Leftist," "Alarmists." Say that ten times fast… On second thought, don't bother; they are all the same thing in my Wikipedia. One could throw in "Liberal Democrats" as well. They are the political force behind the environmental "Green" movement. Since man harnessed fire, environmentalists have been battling against human progress. Why should they stop now that they have a famous figurehead to lead the charge?

Liberals usually give themselves away because they cannot constrain themselves whenever a hot topic is broached. The Democrats (a more subtle subgroup of liberals) on the other hand, are good at misleading you. They are well versed at making you believe they are compassionate and only want the best for you. This often means you have to fork over your money and rights, but it is for the good of everyone else.

Liberal Democrats

Democratic policies indicate that they are more concerned with mandates than individual rights and responsibilities. Liberal democrats are the scariest breed, as they will not always tell you their plans or ideas. They fear scrutiny and debate because emotion is their best friend, not logic or reason. Often, inaction by the government is preferred to activating a bad plan. Unfortunately, inactivity is not always possible, nor can implemented bad plans be easily dismantled.

Democratic politicians only succeed when they are able to divide Americans. They need victims and villains to sway people to support them. Why else would they pit the poor and working classes against the wealthy and upper classes? Envy and jealousy are powerful emotions.

The liberal recipe is to pour a cup of paranoia, mix in a pinch of conspiracies and continuously stir the pot of discontent until all the listeners believe they can't make it on their own. Voilà, you have created the mob mentality that leads to class warfare and more government intrusion. These are the ingredients elitist like to use before devouring more power.

Now that the table of fear has been set on the political front, we begin to see Congress exploit the global warming concerns. Look at the new gas mileage requirements that the democrats want to force on our auto industry, as if the ***not so big three*** automakers are not suffering enough with all the mounting regulations and pension costs accrued over the years. Let's just allow foreigners to make all the cars…

Should congress dictate what kind of automobile you can purchase? How much energy you use? When does it end? I will tell you – it ends when we stop voting for the cliché, "we must sacrifice in order to save the [planet] or [children] or [old folks]." Just fill in the brackets with your politician's favorite group.

As long as there are perceived injustices (someone has more than someone else), democrats in congress will sell you on their plan to level the playing field. How? With more taxes and regulations! This does little more than increase their program budgets, line their own pockets with more campaign pledges, and makes more people dependent on them. Power begets more power.

It is unfortunate that one whole party completely gave up faith in the American individual and the potential for success that freedom and self determination brings. Since JFK, the democrats have given up on promoting a real positive or optimistic view of our future. Next time you listen to a democratic speech, debate or comment, take note of the tone, the negativity and the condescending message.

"Freedom is our ally, totalitarianism threatens all…"
Unknown

Government vs. Private Sector

It is hard to find a government program that has been completely discontinued. It's hard to find one that has ever had a budget cut. When a conservative attempts to just slow the growth in any of the many redundant programs, he is accused of cutting funding for that program. That is not a bad thing, so why do leaders hesitate to cut budgets? It's because there are too many dependants who vote with their hands held out and government workers with no interest in working the private sector, both with tears in their eyes demanding the rest of us pay up!

One time I watched an interview with a local representative (long time Congresswoman), which was as painful as a toothache. She was asked to name a successful government social program. Her answer was a good one, but made me realize an important lesson regarding government efficiency. Play along at home and take one minute to consider your answer to the question, "Name a successful government social program?"

The long time New York State Congresswoman's answer: The GI Bill. It is entirely true that this was a terrific program for GIs to get their degrees after fighting the Vietnam War - I mean conflict. I am sure you will agree. However, this program was over 35 years ago!! She had to go back that long to recollect a successful government social program! If government ran IBM, computers would still use vacuum tubes. If that does not tell you something about the inefficiency of the government, I am not sure what will.

Our government's role is to protect peoples' rights, promote free trade and to protect us against foreign enemies (refer to the US Constitution). Little by little, we have allowed the government to encroach into our business to the point now that the Congress is regulating the size of our toilet bowls (see HR 623). I could write a whole book on government waste, and I just might. When do enough Americans believe that this and other government mandates have crossed the line? Perhaps we already have (Google TEA Party movement).

Many books and many calls have been made to "take our country back" from big government subverted by big lobbyists. Unfortunately, I only see the number of government employees and Czars snowballing into a huge bureaucratic avalanche. Until we decrease people's dependency on the government (i.e., increase individual responsibility), I fear the government will continue to storm out of control.

The main reason the private sector is more efficient than the government is because the private sector can react quicker, since their very existence depends upon it. Like a species struggling to survive, businesses have to adapt, manage costs, and satisfy customers while solving a need. Profit motivation is a powerful thing.

Unfortunately, our government does not have that kind of motivation. It is often hard to find any customer service incentives in the bureaucratic workforce we call civil service.

Once in a government job, it is difficult to get ousted. Contrarily, if you do not produce in the private sector (barring unions), you most likely will find yourself in the unemployment line.

What often happens to a successful business that grows too large? It becomes inefficient. The same thing happens with big government. If allowed to happen, departments and staff enlarge to the point of no return. It is the nature of the beast, so be careful what you vote for.

It sounds capitalistic to talk about free enterprise solving problems. I point this out because liberals hate it when businesses solve problems. To them, that is the government's job! Ironically, that is what our nation does well; we solve problems.

Americans are good at addressing needs and wants when allowed to make a profit from their accomplishments. When is the last time a government agency sent you a satisfaction survey? Even if they did, it would not change anything because they get paid regardless of their ability to serve. On the other hand, the private sector constantly monitors its customers' satisfaction because they have a vested interest in keeping their customers happy. Without their patronage, the company goes under. The government agencies just don't seem to have the same motivation.

There are two sides of this ideology. It is my contention that the private sector is much better at solving problems and filling needs than the government sector. If you agree, welcome to the conservative side. If you prefer paying higher taxes and endorse more regulations, regrettably, that places you on the liberal side.

Teach a Man to Fish

Liberal solutions are often based on good intentions rather than good logic. This is why I assert that the liberal ideology is dangerous: partisan political intentions usually result in unintended consequences. When basing your social programs on what is perceived as good intentions without careful consideration of the ramifications, often lead to undesired outcomes. If liberals did consider their actions or were held accountable for their policies, they would be out of business -- period!

Here is an example of the liberal mindset of good intentions without thinking through the consequences: *Feed all the starving people in the world.* That seems like an altruistic humanitarian goal right? Okay, let's think this one through. We actually could produce enough food (for awhile) so everyone on the planet gets three square meals a day. Let's assume the liberals push through a bill mandating we feed the starving peoples of the world. What would the consequences of this action be?

First, we would have to subsidize farmers to farm all their land and force them to donate a good portion of their crop. The subsidies would come from higher taxes, of course. Billions of dollars would be needed just for the transportation of the food. Can you imagine how many ships, trucks and planes would be required? Many places around the hungry world are so remote that they have no roads. What can *Brown* do for them?

Also, someone somewhere will inevitably be overlooked, causing more angst and outrage. I mean what would you think if your cross-town rivals get free food while you sit there and eat dirt? Question - how do you determine who doesn't have enough to eat? They must already have <u>some</u> food, or they would be dead. How much is enough? Is it possible to provide a balanced meal? Can we get dessert with those fries? I don't mean to slight the hungry people of the world, but I need to point out the failings of the liberal thought process.

The logistics alone are staggering considering the transportation, warehouses and packaging required, not to mention the complete cooperation of all the different governments (some hostile). It's just not possible to feed millions of far away people, and even if it was, it still leads to newer and bigger problems...

Let's pretend the liberals were able to somehow bilk Americans into this project; it would only be a matter of time before our resources (farming and financial) were used up, only to leave more dependent communities stranded. What would they think of America after the free food stopped dropping in? This program would come to a disastrous end once we are all taxed to the hilt, cannot afford our own food, and we join all those poor hungry peoples of the world. Hey, but the liberals' intentions were good, right?

We are talking Economics 101, but think about how many benevolent people liberals could convince to support this plan? Too many, I believe. Grandiose ideas like this and universal health care for all scare the hell out of me. Why, because they are gift wrapped with good intentions, but contain dreadful consequences for all of us. If you don't believe me, read the next section about the war on poverty or, better yet, just think through the consequences of taking such drastic actions...

One more important human trait to consider: give people something, and they tend to want more. If we met everyone's sustenance needs, wouldn't they propagate? A population explosion would only amplify the problem exponentially, right? Why would poor people bother trying to farm for themselves if the food is rolling in for free? Can I get a school voucher with that sandwich? Vote yes to bring back public housing projects?

Isn't free health care important, too? When does it stop? The handouts would go on and on...

The good news is that America does more to help people in this world than any country has ever done in the history of mankind, period! The dollars and time donated are incalculable. We have so many charities, public and private, that provide relief for so many people and countries it warms my heart. So, take pride in the fact that we Americans do already help millions of people, and will continue to give without government mandates.

FYI, do you know what our government considers someone in America as starving? Because there are too few, the government rephrased the terminology to be "low food security."[6] Isn't that clever considering obesity is at epidemic proportions. Where are these 'hungry' people? Is being hungry every now and then necessarily a bad thing? How is 'hunger' classified? Simply as not feeling full all the time? Really--just what does this mean? It's ridiculous that the government has to lower the standards so far to enable them to continue unneeded welfare programs. I say that the government should experience "low tax security" and go without a tax payment until it loses twenty billion pounds of pork!

War on Poverty

You could argue that **the *liberal solution*** suggested above to feeding the world's hungry is a hypothetical scenario. Okay, so let's look at a real live liberal scenario that so few want to talk about. Why doesn't anyone want to discuss the war on poverty? Perhaps this is due to the fact that this policy has been the single biggest failure of modern times: some four trillion dollars in transfer of wealth mandated by government over the last 40+ years and counting. [7]

This brilliant, compassionately liberal policy born in the 1960's was riddled with flaws; but it was enacted with such good intentions, right? Like Moses, we the government will lead thy people out of poverty! Who doesn't want to stomp out poverty forever? Why, oh why, do we still have the same percentage of poor people in America with the same old policies and programs still in place?

I had thought food stamps went away years ago until I read that Nancy Pelosi was trying to tie an increase for one of the food stamp programs with the 2008 stimulus package. I guess these programs will never die as long as one child misses snack time. It never occurred to Nancy that if we cut these programs, allowing her to cut taxes, we all would be able to afford the food again.

Unfortunately for mankind, there has always been a poor populace. Ever since the first human was labeled wealthy, everyone beneath his means was considered poor by default. The only way to extinguish poverty would be if everyone everywhere had the exact same *stuff*.

So, instead of raising everyone's chances at attaining the same *stuff*, the liberal solution was and still is to take *stuff* from the people deemed wealthy and give it to the people labeled poor for free so long as they cast their vote their way. This also was the origin of the phrase "Life is not fair," or was it "Nothing is certain but death and taxes?" No, I remember, it is the definition of the word **extortion**.

In the 1960's and with noble intentions, LBJ and his clones enacted the infamous *war on poverty*. (Not sure if he was nominated for the Nobel Peace Prize for that one.) Although everyone had food, some had more than others. Contrary to human nature, liberals believe that we should all eat the same whether we earned it or not!

The government's main obstacles to converting us to complete socialism are those darn selfish achievers who want to control for themselves which charities they donate to. This is basically what we did before governmental socialist policies took root; forcing achievers to donate more tax dollars so the government can pick and choose who gets what is by definition Socialism. For those too young or uninclined to read history, people used to take care of people. Family fended for family. It was not the government's role until the liberal's hero FDR began ruling. Since then it has snowballed out of control. Stop and think about what kind of people want to control a trillion dollar budget and still think it is not enough?

The Surge Against Poverty

The second part of the war on poverty, "the surge," is the logistical component, and it turns out to be the hardest part to understand. The concept seems heavenly, but the devil is in the details. It turns out that the "haves" are a voting minority (perhaps 10-20%, depending on your definition of wealthy), and are limited in their numbers to effectively defeat these initiatives at the ballot box. My mom used to say, "A loving family and good friends made you wealthy, not money."

Try asking your nearest democratic politician what their definition of wealthy is and if they don't get hung up on the definition of what "is" is, you will be surprised. Remember, Al Gore Jr. during his bid for El Presidente, defined wealthy as anyone making over 36k a year. So, if you are not in the top 10% and you do not oppose legalized extortion (i.e., income taxes), it is easy for liberals to pass more punitive tax rates. As long as taxpayers are the minority, the majority of "recipients" will continue to vote for more spending and more taxation.

Since liberals have already convinced the middle class they won't be affected by higher taxes (in fact they have been so bold as to promise tax cuts to the middle class), those liberal politicians will continue to be in the coveted majority.

The middle class is always affected by higher taxes because if you make any kind of money, the government goes after it somehow, someway. Getting back to extortion (I mean transfer of wealth, oops; I mean the compassionate war on poverty), how does a liberal politician transfer all this newfound wealth to the newly baptized voting poor?

First, the liberals needed a full-scale bureaucracy, then a new wing on every city's government building, complete with a legion of non-conservative social workers. Better yet, let's create a whole governmental department to completely entrench the program. In case we actually do wipe out poverty, we can always create more victims; or, like today, just raise the poverty line higher to continue supporting the request for more and more tax dollars.

Does anyone have any idea what is considered poor in America nowadays? Only two televisions, one car, one microwave, no plasma screen TV... the list is too sad to continue. Are there real Americans in real need? Of course, but do the government agencies accurately seek out and find those justifiably in need? Who decides what is *justifiably in need* anyway? I guess it is like pornography -- I know it when I see it.

Today we have over 90 federal programs to combat poverty (with poverty being redefined every year). What if we declared victory on the war on poverty? Would the agencies be disbanded? Where would all the social workers go? Wouldn't they become poor and then qualify for assistance too?

Who decides how to best spend what's left of the tax dollars? Who deems someone worthy to receive such gifts? How much should each lucky recipient get? If families do not receive enough aid, they will stay below the poverty line. If they receive too much, they won't ever earn a living themselves. The logistics get even foggier when you consider the demographics: a poor family in Arkansas needs less than a poor family in NYC. Do you give cash or credit? Real food or just stamps? How do you know if they are actually poor? Did you check with their accountant?

Do you know how much of every tax dollar confiscated in the name of the poor actually goes to helping the poor? How much goes to fuel the agencies and the bureaucrats' self-serving overhead? Perhaps I am being too harsh, as there must be some civil servants with genuine intentions. During this renaissance, private charities (including faith-based charities), which used to care for the needy, were unceremoniously bumped out by larger government mandated programs. I would like to see a revival of purely charitable organizations retake the ominous task of helping the helpless and leave the bloated, politically motivated government out of the equation.

Government social engineering is not a hypothetical example, but a real life objective our liberal counterparts are determined to win at all costs. To illustrate, in over 40 years of the so-called war on poverty, our government has redistributed trillions of dollars from taxpayers to the so-called poor. However, the percent of impoverished remains about the same; results don't mater, remember, just good intentions.

In a real war, we could measure success by how much territory we controlled or by how many enemies we killed. In the war on poverty, it should be easy to count the number of people that were in poverty, however defined, and after receiving government assistance are now out of poverty. That should be easy to measure, right? Well, as you might have thought, we still have the same percentage of poor people now as we did in the 60's. It is as if we sent our troops into battle for forty years and never gained an inch of ground! Why hasn't Harry Reid waved the white flag on that lost cause?

Are you done screaming yet? How could this be? How in God's green Earth could we lose the war on poverty? Simply stated, good intentions do not always translate into good results. Unfortunately, in our rush to win, we forgot to think this through. The fact is when you give people a handout for extended periods of time, they become dependent and dependency stifles motivation.

Lack of motivation is often what keeps people down and out. I could easily list a myriad of successful people who were once qualified as poor, but willed their way up and now live the American dream. Don't be fooled. Self-respect and self-achievement go a long way; or, dare I say, will take you a long way down the road to success, no matter how you define it...

The Traditional Family

No society in history has become successful without the strength of the traditional family structure. Check it out. It's true. In fact, the decline of successful societies coincides with declines in the family structure. I am not going to play the part of a *doomsdayer*, but the decline of the family unit here in America is disconcerting to say the least. The cost of dysfunctional families may be difficult to calculate, but it is not difficult to correlate that trend with other societal problems that happened to increase at the same time.

Looking back through the 60's, it almost appears as if liberals wanted a breakdown in the family. I mean, if I were attempting to subvert the traditional family structure, I would do exactly what the liberals did. First and foremost get as many people dependent on the government at every level (e.g., welfare). Never encourage people to rise above their circumstances, as they are really just victims, right? Then create a system where single moms receive more subsidies than married poor folk. Throw in a little disincentives (i.e., high taxes) for attempting to make a little money, and what do you think happens to the family structure?

Does it seem like the government was paying all those poor single moms not to get married (get more benefits), not to work (decreasing benefits) and to have more children (increasing benefits)? Who benefits from an impoverished, dependent, discontented, angry group of people? Answer that, and you see the real motivation hidden behind the good intentions…

Can you guess what percentage of the U.S. poor vote for Democrats? If you were dependent on the government for your meal check, which candidate would you favor: the candidate promising smaller government and fewer taxes or the one touting more social programs and raising more taxes to fund them? The democrats plan to win the war on poverty, in reality, has done nothing more than treat people as sheep, sheering their votes and keeping them penned into poverty. Meanwhile the shepherds maintain their majority in congress, and the conservatives are labeled as wolves for disagreeing with the status quo.

I just want to say a word about population growth here, because it goes to the crux of the issue regarding greenhouse gases. You could argue that having fewer people on the planet could reduce all kinds of epidemics and other global issues. Though this is true, it is a perplexing issue to say the least. How does a global community agree to reduce its populations? It is in each country's own interest to increase its population for various reasons like defense, labor, and diversity.

Are growing populations of humans with growing needs for resources savaging our ecosystems worldwide? Yes! Should we come up with a plan to curb the world's population explosion? Certainly, I am all for reducing the population in China, Pakistan and Iran, as they are the world's biggest threat to peace and prosperity. I omitted North Korea because it's already cooperating by reducing its populace (sadly through starvation and/or neglect). So put together a plan to control worldwide population in these places, and I will help you sell it.

I am not aware of any country ever successfully reducing its population. Yes, China has tried to limit the number of offspring the commoners can produce, but it is far from actually reducing the population. We may not be at critical mass yet, but someday we will have to face the dilemma. Unfortunately, the most effective methods of population *reduction* are starvation, disease, war, and yes, I almost forgot, abortion. As you can see from the list, these options are neither popular nor humane.

The New Tone

No president since George Washington has tried harder than George W. Bush to bring civility back into politics. The education bill, prescription drugs bill, and the amnesty bill were all right out of the liberal mindset. Did that unite the *donkeys* and *elephants*? Hardly, comments like "the President did not go far enough..." and "It's just a start, we need to demand more," rang out within days of drafting these bipartisan bills.

Nothing is ever enough for liberal politicians. If 70% tax rates weren't high enough, how could a trillion dollar prescription drug bill be enough? As long as liberals believe that achievers (i.e., producers) at every level should be taxed into oblivion and the proceeds doled back out to the Democrat-chosen non-producers (pc for under achievers), then, my friends, the right and left will always be contentious.

Ask a liberal what percentage of the wealthiest 10% earned their money and what percent inherited theirs. Most will say 80% inherited and 20% earned. This fallacy is perpetuated in the liberal media and implied by liberals everywhere. The fact is just the opposite -- most wealthy people have indeed earned their fortunes and continue to be producers (i.e., job creators) in our country. The other 20% includes the infamous Paris Hilton, which does give the rich a bad rep.

If you have not heard this statistic, you should - the top wealthiest 10% pay 30% of the federal income taxes in this country. Why then would liberals want to crucify the very people who keep this country afloat? It is my contention that more and more liberals want others to have less affluence. Their policies attack producers and achievers while rewarding the stagnant. This fits well with the liberal contention that global warming is caused by man-made activity. If you eliminate the activity, the global warming issue goes away. The problem with that solution is that if you decrease the producers and increase the non-producers, which way do you think this country will head? It seems obvious to me what does your common sense tell you...

Liberals rarely propose viable solutions. What is their solution to our oil dependence? Rope off the frozen tundra in Alaska and conserve, conserve, conserve! You cannot conserve your way out of this problem; it may prolong it, but it certainly will not solve it. No my friends, liberals are much better at complaining and finger pointing than finding real solutions.

They shout, "Conservative proposals benefit big business!" Well, the reality is that by giving businesses incentives (i.e., tax credits) to solve a major problem, viable free market solutions do present themselves. Free market solutions benefit everyone suffering from that problem, right? Otherwise, no one would buy his or her solution. If a business cannot benefit (i.e., profit) from its own solutions, then we no longer have a free market system. Why go to work if you are not going to get paid? (I know some communists that would like to answer that rhetorical question.)

Stifling real solutions creates more and bigger problems. If government actually solved problems, we would not need so many agencies, right? Of course, they perversely benefit from not solving problems until they are so massive, people feel like the government is the only entity big enough to save the day.

Global warming advocates look to big government in vain to solve their perceived problem. If they only realized that the government, as big and strong as it is, cannot fix this problem any better than it fixed poverty. The only gain would be that big brother would have another link in the chain around our neck.

Again, government does not have the incentive to solve problems; in fact, the liberals benefit the most by allowing the problems to grow to monumental proportions, at which point the liberals mandate to create another program or write another billion dollar check. Who benefits again? You guessed it, liberal politicians and lifelong bureaucrats.

The dirty little secret behind welfare reform was that the government handouts had not kept up with the latest expanding economic potential for even the least lucky Americans (legal or illegal). More economic progress (thank you, tax cuts) led the way to the lowest unemployment rates in history. The underprivileged found jobs and/or business opportunities more favorable than standing in the welfare line for their meager payoff. How do you suppose welfare reform ever passed?

More and more people that were tricked into the welfare system discovered they could make more money in the real world of capitalism. With increasing bureaucratic red tape to endure in order to keep the measly portions doled out by the social welfare agencies, it actually became easier (and more rewarding) to work a 9-5 shift than to stand in the welfare line. How ironic -- the welfare state is collapsing under its own weight. Yes, there is an obesity problem, and it is not just confined to American citizens...

"Stupid is forever, ignorance can be fixed. "
Don Wood

Chapter 8   Save Our Planet

Some people just need to feel important. Others have a desire to control. Ask a liberal why they chose their profession, and most likely they will say it was so they could change the world. Don't get me wrong, changing the world could be a great thing; technological advances, medical cures, communication and transportation all have eased suffering and advanced mankind. In contrast, the global warming hucksters have done nothing progressive; unless you call frightening women and children into believing their future is bleak an accomplishment.

Emotions over Reason

Fear is a very powerful motivator, second only to love; it can generate a powerful action, but also can create an even bigger reaction. Once this global warming con is uncloaked, the masses will finally revolt against the liberal elitist power grab. It is my hope that thinking Americans will become empowered enough to finally act. Pulling a lever every couple of years is not enough. We must consolidate, join like minded organizations, support with time and/or money those candidates and organizations who agree that a smaller government is a better government.

Have you ever watched an astounding magic trick and feel the bewilderment, only later to discover the trick's simple ploy and how that made you feel almost ashamed? Don't fall for another leftist ruse. Look behind the smoke and mirrors and see the truth – the sky is not falling. Please don't live in fear or self imprisonment over this global hoax. Let reason and common sense rule the day. I do.

Remember, liberals usually have good intentions regardless of how bad their program's collateral damage is. Look at our welfare programs over the last few decades; that was four trillion dollars well spent. Granted, the intention of helping the indigent is a noble cause, it's just that the irreconcilable consequences of broken families and insufferable dependence the programs caused rarely get any attention.

Liberals have an incessant need to tell others what to do and how best to do it. It's the unapologetic know-it-all club. It is also one of the most intolerant and close-minded groups I know and ironically that is exactly what they claim the right is. On the contrary, they will argue their innocence until they are red in the face. Sorry, liberals prefer the term "blue in the face."

Why do I slam liberals so hard? Am I that agitated? Yes! I get agitated when a group of busybodies tells the American people that they are causing the Earth's impending doom, and they must use less, do less and give up more of their wealth or else! Nowhere in their rhetoric is there any confidence in American ingenuity or American individualism; at no time do they recommend we should rely on common sense solutions. Government mandates seem to be their only answer. If they are right everyone would just follow their suggestions, but they are not, so they have to make it a criminal act to use an incandescent light bulb.

Since Ida's son, how many hurdles has mankind overcome? How many American inventions have improved the world? How is it that 200 years of independence and a free market economy has created the most successful civilization ever? Why not promote solutions instead of requiring sacrifice? Why not advancement over digression? How about hope over despair? I get agitated when an ex-Veep cannot see the abilities and possibilities in his own countryman. Perhaps he does not have faith in Americans.

Liberals have a powerful desire to control others; either be in control (e.g., legislators) or have like-minded people controlling others (e.g., bureaucrats). They are not happy allowing people to make their own decisions, because people may make decisions that offend others. Look at the schools that are predominantly run by liberals. They are the most P.C., intolerant, pretentiously run establishments around. They get to rewrite history, select the facts, and mold young minds of mush to their leftist agendas. Tradition, culture, respect and truth lie in detention as the liberal agenda continues onward and downward.

Liberals would have you believe that a strong central government is the solution, even though they cannot point to a single government program that has overachieved and came in under budget. On the other hand, I cannot count the multitudes of American companies who have become hugely successful because they discovered ways to solve people's problems at a cost they can afford. They are just too numerous to list (Perhaps that is a whole other book in itself).

Can you name a successful governmental program that was so successful it is no longer needed? Anyone? No, I didn't think so, because governments only grow. Even the wars we wage cost too much and last too long, as we all painfully know. Regardless of politics, everyone should be thankful we have strong brave men and women willing to fight for freedom.

As a citizen and a consumer, which entity do you think **you** could effect change upon: a government bureaucracy or a profit-driven business? The dirty little secret is that the government agency will be funded whether you are satisfied or not. The business, on the other hand, is successful only if it satisfies its customers. When is the last time the government sent you a satisfaction survey?

"The fewer the needs of the people, the less the need for government."
Unknown

Left vs. Right Solutions

Ideology

Conservative   vs.   Liberal

Here is a simple concise break down of the differences between the conservative and liberal mindsets. This is not my opinion; this is extracted from actual positions and actions taken by each group.

Issue	Conservatives	Liberals
Environmental Stewardship:	Personal responsibility and incentives	Government control & punitive damages
Government's Role:	Limited	Pervasive
Political Strategy:	United we stand	Divide and conquer
Policies/Solutions:	Long Term	Short term & short sighted
Concentration of Power:	With Individuals	With Government
Mankind's Future:	Optimistic	Pessimistic

The liberal solutions versus the conservative approaches to global warming and most every other issue could not be more dissimilar. On the left hand, we are to put the fate of the entire world (the future according to Gore) on a bloated bureaucracy, beholden politicians and liberal judges. On the right hand, we conservatives would rather put our faith and money on private enterprise and free markets to solve these issues.

I have no problem with government incentives to attack certain issues like pollution or even some research, but generally if there is a genuine need, entrepreneurship will step in with free market solutions. There is good money in viable solutions, as there should be.

There is little motivation for government bureaucrats, as they never work on commission nor do they get bonuses for exceptional work (excluding kick-backs). So who would you put your stock in: a civil servant with job security not based on performance, **or** a private entrepreneur who is goal oriented, profit motivated and a completely accountable problem solver???

I'm certain someone reading the previous Ideology table will scoff at the suggestion that conservatives are uniters and liberals are the true dividers. I realize every day we here from the left that the opposite is true. I take exception to this tactic of blaming the other side while continuously employing the very tactics that do divide this nation. Politics are dividing our country and government is a behemoth enterprise now, but it is a false notion that conservatism is causing this division.

People should be good stewards of the land, water and air. We should stay vigilant against all polluters and give tax credits (i.e., incentives) to invest in new "clean" technologies. Conservation is a good thing as long as it does not jeopardize our liberties and freedoms. We cross the line when we force out all our factories and send the manufacturing over to third world countries that do not have the technology or the means to minimize pollutants. We would be better served having all the manufacturing here in this country. Why? Because, here in the good ole U. S. of A., we have the resources, the technology and the brains to keep our environment safe while pursuing a better standard of living for everyone.

Carbon Rationing

The concept of carbon rationing is simply wrong. Since the gasoline-rationing program of Jimmy Carter era, no rationing program ever devised has worked. Every human on the planet exhales carbon. Are we to ration everyone's breathing? Of course not! Government officials, the media and Hollywood stars will be exempt. Even if we wanted to limit some people's emissions, who will get to decide? Would you feel comfortable handing over that immense power to a politician with an agenda and contributors wanting pay back for getting them elected? Sorry, the truth is government officials are not as altruistic as liberals would have you believe.

If the government cannot ration our breathing (another CO_2 emission) they surely can tax it. It was not too long ago our fearless leaders wanted a BTU tax. That was a proposed tax on basically all energy usage. Really, we don't pay enough already for energy? Who sits around and thinks up all these new ways to tax us? I'd like that job; is it based on commission?

If we correctly think it is wrong for the government to ration and micro-manage all emissions here in the US, why would we think it could be done on a global scale? More importantly, can it be done fairly? There is a 0.001 percent chance (rounded up) at getting every nation to agree on a single fair plan. This is not pessimism, it is based on the fact that every country must look out for its own best interests first, and what is fair for us may not seem fair to them.

Furthermore, poor nations will whine they don't have the resources to cut back emissions, while rich nations will argue their having to burden the brunt of the restrictions. The tripe goes on and on; if you don't believe me, sit in on a UN meeting sometime. The only thing those UN blockheads agree on is that no one should get punished for stealing billions from the Food-For- Oil programs.

Supply and demand has worked well for this country in the past, and I believe it works better than any body of politicians who think they know what's best for everyone else. Administrations come and administrations go, but businesses are in it for the long haul and are much better at planning ahead.

How the government has been able to infiltrate our lives is a sad state of affairs. It is a path to socialism, which has not proved to be a viable or fair governing system anywhere in the world. The only fair system I have seen is the one where individual rights are respected above governmental powers; where people are allowed to participate in free commerce; where citizens are allowed to prosper from the fruits of their own labor. However, with individual rights comes individual responsibilities, and sadly, that is where some of the people have faltered.

It appears to be a daunting task -- to borrow a phrase, "You cannot fight city hall." But, everyone can contribute something to the process of remaining a free union: Step 1 - vote. Vote for candidates not based on party affiliation or name recognition. Instead, vote for candidates who respect peoples' freedoms over government interference. Vote for honest candidates who promote smaller government, fewer taxes and, most importantly, believe in American exceptionalism!

"I have faith in my brothers, just not in big brother."
Unknown

The Jiffy Pop Solution

A rush to biofuels may be a mistake and, perhaps, converting a portion of our food supply to make a fuel additive is already causing greater problems. I like the idea of supplementing our gas with a homegrown additive in an effort to reduce our need for foreign oil. However, like many liberal ideas that start with grand intentions, this is beginning to strain other commodity prices.

Over-reaction to global warming is bipartisan. President Bush proposed that we replace twenty percent of our current gasoline consumption with ethanol over the next decade. But it's well known that even if we turned every kernel of American corn into ethanol, it would displace only twelve percent of our annual gasoline consumption. The effect on global warming, like Kyoto, would be too small to measure. The U.S. would become the first nation in history to burn up its food supply to please a political mob.

When you remove supplies of a major commodity like corn from the market, buyers pay higher prices for what is left. We are seeing the effects of less corn by rising prices for animal feed and other consumer products. I'm sorry to report that any savings we may see at the pump will only offset higher prices at the grocery store.

Good intentions gone awry are generally the result of *liberal reactionism*. It may sound like a made up term, but it's not. Every action has a reaction, so we need to consider carefully the consequences no matter how good the intentions (remember the feed the hungry scenario). In this case, we need to examine the consequences of proposed reductions in greenhouse emission. They could cause more problems than they purport to solve.

Who has a better chance at solving global issues: wealthy and technologically advanced nations or third world countries struggling to feed their people? Policies that reduce our ability to produce (i.e., make money) curb our ability to invest in real, viable solutions. In other words, a rush to overtax and over regulate our industries will lead to less capital for research and implementation of meaningful improvements.

The growing trend to offshore work from Americans is a consequence of big government, not greedy CEO's. I don't know a single CEO who wants to push work offshore if they were not forced to do so. The cost of doing business in America with American employees has outpaced what other countries' laborers are willing to charge.

When you add up the FICA, Medicare, SSI, and a myriad of other costs forced onto employers, I am not surprised that they are willing to bid on outside labor whenever possible. When we make it more affordable to hire Americans, more Americans will get hired. Now that you know this, let your congressman or congresswoman know, too.

"Change for the sake of change may only change the problem."
Unknown

I'm Scared

It <u>does</u> scare me when top-level leaders begin to jump on the global warming bandwagon. This is a prelude to more government action, which will most likely do more harm than good. The proposal to start banning products like incandescent light bulbs is out of the realm free people expect from their government; let the market drive life style changes. Politicians should stick to research and education if they are compelled to act, but don't legislate a free society with a free market system what kind of light bulbs are permissible!

Sympathetically, I don't get scared when businesses begin marketing green products. I could argue that it only perpetuates the hoax, but that is not consistent with capitalism. Every business has the right to market their products in their own way, as long as it is not misleading. If GE wants to tout their products as environmentally friendly, fine, as long as they are, go right ahead. At least businesses are not making it a crime if I chose to buy a competitor's incandescent bulbs.

My goal from page one was to reason with you, plead with you to think, and help you to understand that this planet is not faltering us. Be aware that a small group of control freaks and hustlers are attempting to scare you out of your money and lifestyle. Don't let them do it; don't let them take our freedoms!

If a hundred years ago the media touted that the Earth's temperature would rise one to two degrees over the next hundred years, what would you have done then? I know I would have said, "So what, I will be dead in a hundred years." Just to make sure my grandkids wouldn't panic, I'd teach them to be hardy enough not to cringe or panic at a one to two degree temperature change. I would tell em, "Go invent air conditioning and become a millionaire!"

Seriously, nowhere do I discourage conservation, prudence, or responsibility. We all benefit from conserving, protecting, and taking responsibility for our environment. I don't drive a Prius, but I do have two recycle bins. Please don't take this book as a justification for pollution or wastefulness. Just as we should drink and drive responsibly, we should heat and energize responsibly, too.

Ideal Temperature?

What is the ideal temperature anyway? Seventy-two degrees Fahrenheit? Partly cloudy? Partly sunny? Who should the weather be ideal for: people, animals or plants? Don't they all vary? Interesting question, since some people are now claiming that a variation in yesterday's mean average temperature is now somehow intolerable.

There is no reason to believe that slightly lower temperatures are somehow preferable to slightly higher temperatures. There is no known *optimal* temperature nor any means of knowingly and predictably adjusting some sort of planetary thermostat. Fluctuations in atmospheric carbon dioxide are of little relevance in the short to medium term.

Where I grew up people, plants and animals adapted to changing temperatures quite nicely. I bought a heavy coat for the cold northeastern winter and a light jacket for the milder spring and fall with t-shirts optional for summers. Plants seem to come back every spring along with all the animals, and they don't even have the benefit of a clean-burning, 95 % efficient natural gas furnace. How is it possible for them to survive, with such a fragile ecosystem, under-matched by all the evil factories?

I am certain the peoples of the republic of Siberia are anxiously awaiting the prospects of a one to two degree spike in their ideal temperatures. I even heard the citizens of Buffalo, New York, were planning a December parade to celebrate the coming of the excess heat. I think they are calling it "no more frostbite" day.

Perhaps the Inuit's are concerned their ice huts will melt, but you know somewhere there is suffering in the world; there always has been and will always be. That fact doesn't need a culprit, nor does suffering mean victimization. Everyone everywhere has to grow, adapt and overcome hurdles to ensure prosperity or sometimes-just existence.

Do the alarmists actually think that we will believe that mankind is so fragile we could not overcome a minor climate fluctuation? Whether you believe mankind has been around for ten thousand years or half a million years, it certainly has endured and yes, thrived through much tougher climates than this.

In many areas the average temperature varies tremendously. It is not that uncommon for the temperature to rise and fall 20+ degrees in one day. From morning to midday to evening, the temperature can and does change wildly. To even argue a .5-degree will alter an entire ecosystem is ludicrous. I cannot detect a .5-degree temperature change indoors or out, can you?

I know a few guys who can detect a one to two degree temperature change in their beer, but that took years of evolution to hone that skill. That gives me confidence that the majority of humans can adapt if they had to. Liberals on the other hand seem to stay their same predictable selves. Too bad they are missing out on a wonderful once and a life time opportunity called life.

Most inhabited areas deal with large temperature variations every sunrise through every sunset! In fact, all inhabitants are well adapted to some temperature changes. Remember the section titled "Desert Life" about inhabitants that thrive on temperature swings. This planet does not just tolerate change; it works to benefit from a myriad of climate conditions. From deserts to frozen tundra, from swamps to forest, these ecosystems don't just endure; no, they endeavor to flourish.

Furthermore, every creature, including man, that was able to adapt to its surroundings, overcame climate challenges or died out. Fact - the strong have endured to propagate over the years. To assert a one to two degree temperature increase will destroy this highly evolved ecosystem is pure folly! Perhaps some humans have not kept pace with our genetic progress, but they are the minority to be sure…

"Common sense is instinct, and enough of it is genius. "
Josh Billings

"Nothing is more fairly distributed than common sense: no one thinks he needs more of it than he already has."
Descartes

Chapter 9 The Future Begins Today

One of my all time favorite movie lines is from the Star Trek series where the head alien who wants to enslave humans into the Borg collective tells our Captain, "Resistance is futile." I like the quote not because resistance is futile, to the contrary, resistance to enslavement is noble. Our American foundation was built on the mantra, "Give me liberty or give me death." Our willingness to resist authoritarian control is what sets us apart.

Some reputable scientists who have attempted to disagree with the "collective" have been ostracized and/or banned from the elitist scientific community. As more shenanigans (Emails) are uncovered, we common folk will see the true nature of their bias. Is it possible that the millions in grant money have not influenced their conclusions at all?

Trust me when I say it is a religion to the extremists supporting the theory of Man-Made Global Warming. Like religion, it requires faith -- faith in the notion that man is causing a rapid, worldwide destruction of planet Earth. Be on guard as we are entering the inquisition phase of the global warming zealotry, seemingly, this movement is becoming as intolerant and righteous as the 15th century Spanish Monarchs. History does repeat itself and fortunately we have the ability to learn from it.

It would be nice if the experts would actually level with politicians about various "solutions" for climate change. For example, the Kyoto Protocol, if fulfilled by every signatory, would reduce global warming by 0.07 degrees Celsius per half-century. That's too small to measure, because the Earth's temperature varies by more than that from year to year.

However, what is not hard to measure is the enormous amount of wealth that would be sucked out of Americans and given to third world regimes. A measure of insanity is to trust 3rd world dictators with millions of taxpayer dollars. How do you think they got to be 3rd world countries?

Victory is at Hand

One glorious victory against the anti-human, doomsayers' crowd came one day not long ago from the Glenn Beck radio show. Glenn shared how a 5th grade science class ripped the Al Gore Jr.'s dying planet film to shreds. You have no idea how much faith and hope that gave me. If 5th graders could see through the hype and disinformation, then what is the chances average Americans with an iota of common sense and rationale could also see through the bunk?

If it weren't for the incessant barrage of doom and gloom from the liberal media and some so-called leaders selling out to this aberration, I think it would have passed like a jagged kidney stone months ago. The debate might be over, but the fight for the hearts and minds of reasonable people will continue on, at least until Al Gore's 747 chokes the life out of us.

Another ancillary reason why many Americans see this theory as implausible is simply because the so-called effects will take a predicted 50-100 years to reach a critical point. Either people realize they won't be around then or more likely that it is so far away that things could change by then. Change got us into this so-called mess, so who's to say change won't get us out?

According to the latest studies, the global temperature has not risen in fifteen years. How is that possible with all the trains, planes and automobiles scurrying about? Didn't Einstein conclude, "Every action has an equal and opposite reaction."

I believe Americans have faith in our ability to overcome obstacles. We have moved mountains or, when it made sense, bored a hole right through them! We have dealt with worldwide issues like wars and prevailed (until the liberals stifled the army in Vietnam). The people who are scared are the ones without hope or without confidence in American wherewithal. That is sadder than Al Gore's movie, because we have become the greatest country **ever** because we have the bedrocks of faith, hope and freedom.

What about all the recent hurricanes? There has to be a reason why the world endured so many hurricanes in 2006. Why not blame it on global warming (the one-stop-shop for natural disasters)? If it is too cold, it's because the southern hot air is sucking the northern cold air further south. If it is too warm, that one is easy: you left your SUV running, and it created a greenhouse effect, which in turn increased the temperature outside. With this ingenious reasoning, we could all leave our refrigerator doors open a minute a day – that should get the temps back down to normal.

The funny little thing is that while that all sounds plausible (especially when gift wrapped by slow talking southern gentleman and personally delivered to you), logic and reason get in the way. Yes, hurricanes were more plentiful in 2006, but wait - we had a shortage in 2007. Strange, if we are truly getting warmer, this trend should have continued. No need to fret, as it turns out. The weather comes and goes in cycles. It always has and, historically thinking, always will.

No matter what the calamity, it always comes back to one thing - man is destroying the Earth, got it? Now that that is proven, we need a savior; someone who can communicate the problem and spearhead the appropriate steps to reconcile man's incessant consumption. Perhaps a failed politician could fit the bill. Do you know any that are available?

Politicians Gone Wild

Harry Reed, while Senate Majority Leader, has gone too far. In 2007, he stooped so low as to call out a private citizen from the Senate floor. Rush Limbaugh, whether you hate him or love him, is a private citizen, and his views (i.e., lawful free speech) should not be censored by a Senator who disagrees with them. This is a dangerous precedent for a free society. Our freedoms are quietly dwindling, and we must draw a line in the sand before they are irreparably washed away.

Liberals, instead of perpetuating negativism and promoting big government positions, need to rejoin the American way. American exceptionalism that has made this country great was generated from individual responsibility and entrepreneurship. No individuals I know have reached the pinnacle of success via a government program or handout.

It seems Political debates quickly turn into blame games and conspiracy theories, instead of accurate portrayals of policy initiatives. Why? Because most people don't want to pay higher taxes and endure more regulations. If fewer people (currently 1 out of every 4) were not so dependent on the government for their sustenance, we would not have such a divided country.

Politicians are beholden to their constituents' right? Then why do most spend a quarter of their time fund-raising? They need money to stay elected, that's why. No money, no power. Where do elected politicians get the money? From people and businesses with money to buy influence, that's where. It works for both sides. So before you hand elected officials our power, find out where they get their money...

It is obvious the Democrats are not used to being challenged. Their intentions are deemed so pure; they believe that results don't matter. Well, results do matter in the real world. It does matter if your policies lead to a dependent class of Americans. It does matter if people would rather take a handout than earn a living. These things will eventually destroy a country if we all don't start mattering.

Tomorrow, Tomorrow, I Love Ya Tomorrow

Yes the sun will come out tomorrow. Some things you can count on like the Sun rising at dawn and politicians calling for more government intervention. However, only one of these can be stopped. And they can be.

It is difficult for individuals to believe they can make a difference against the bloated political machines which run our state and federal governments. But, it is individuals who start the rallies that influence policy and encourage like-minded candidates to run. If you cannot start a movement (e.g., TEA Party) then perhaps you could join one. Contribute some time or a few bucks, or just start by putting a bumper sticker on your car. I did and it paid off nicely. I put the bumper sticker "Annoy a liberal, work hard and be happy" on my car and it got me out of a speeding ticket. The police officer who pulled me over was so amused by the sticker; he let me go with a warning. You never know how far a little time, money and effort can affect others.

My best friend's son is going to college for film making. I asked him to help me put a video up on YouTube opposing Obama's health care law. Popular videos get thousands of hits. Anyway you can contribute to the cause - fighting the collective. Go for it, freedom needs you more than ever. You will feel better about yourself if you try; I know I did.

How many remember the Y2K scare? Raise your hand if you recall the dire predictions the world was facing at 12:01 Am year 2000. All the stupid computer chips were supposed to fail unleashing hell on Earth. I vividly remember the fear mongers on TV and the clowns selling survival kits sensationalizing the apocalyptic theory and profiting from it (sound familiar). Y2K had most everyone captivated and scared. Tic-tock, tic-tock, then bam, nothing happened. Not even a wrong phone bill.

Uncertainty is a powerful emotion and the so-called experts exploited it every chance they had. Fortunately, the event had a fixed deadline, a make or break moment. Oh how I wish global warming had a definitive deadline. Even if it took ten or twenty years, we could just wait for it: tic-tock, tic-tock, pssss, nothing. Not even a dead sucker fish... Then we could tear up the cap and trade bill and get back to the pursuit of happiness.

Don't Live In Fear

If you are an American, then act like one! Foremost, take care of your family. Be optimistic – it's contagious. Help your community in your spare time. Don't doubt that you are an important part of this great society called America.

The good news is that you still have the right to vote – use it at every election. For the most part, you still have the freedom of speech – use it to inform yourself and others. Let your representatives at all levels of government know how you feel – trust me, they are not mind readers.

The bad news is that not all bad politicians get voted out. While we still have the power to vote, use this power and use it wisely. Ask fellow Americans to vote, and ask them to choose wisely. Share some of the common sense points made in this book. You will be amazed at how intelligent you will sound.

Unfortunately, the likes of Al Gore cannot be voted out again from his self appointed chairman of the new world. This is a distinct advantage for Al, to be self appointed that is. However, we do not have to buy what he is selling. The power of the purse can work against him. Boycott his ridiculous claims and let his sponsors know we are not buying the Global Warming Hoax!

If you read this book and understand the ulterior motives of the people promoting this hoax, then you must understand why I am not scared of man-made global warming. Call out the hypocrites for who they are. Yes, some actually believe the fallacies, but that is because they neglected to see the facts and failed to exert the common sense training we all have instilled in us. Point them to the right sources, talk them through the history of climate changes, and most of all buy them this book!

"Everything in totality must be considered, or issues become skewed and pointless to argue."
Unknown

Greenland glacier 2009

* The real reason the glaciers are melting

Glossary

Carbon Dioxide (CO$_2$) – A colorless, odorless, non-poisonous gas that is a normal part of the ambient air.

Carbon Taxes – A surcharge on the carbon content that discourages the use of fossil fuels and limits everyone's ability to earn a living. Also, a way for governments to steal more $.

Environmental Whackos – Anti-American and/or Anti-Human extremists who wish the world was made up of small communes.

Kyoto Protocol – An international agreement to reduce the wealth of developed countries and provide subsidies to poorer and more polluting countries.

Pleistocene Epoch – A series of ice ages with severe climate changes.

Trace gas – A term used to refer to gases (i.e. greenhouses gases) found in Earth's atmosphere that taken together makes up less than a fraction of a percent of the atmosphere.

Social Liberals – Control Freaks.

Weather – Describes the short-term state of the atmosphere. Weather is not the same as climate.

References

1. http://www.medindia.net/news/Rising-Temperatures-Could-Ring-the-Death-Knell-of-Snowrelated-Activities-in-US-31841-1.htm

2. http://www.policynetwork.net/uploaded/pdf/cc_sd_final.pdf

3. Global Climate Change: Turning the Tide

4. Official British Court Finds 11 Inaccuracies in Al Gore's An Inconvenient Truth, Labels It As Political Propaganda « BUUUUURRRRNING HOT

5. Irrefutable, It-Is-A-Fact Global Warming Causes 'All-Time' High Antarctic Ice.

6. http://www.heritage.org/Research/Welfare/bg1713.cfm

7. /www.freedomworks.org/informed/issues_template.php?issue_id=1657

8. Johnson, C. C. (2005). Making instruction relevant to language minority students at the middle level. *Middle School Journal*, *37*(2), 10-14.

* D. J. Dana (2010). *My Brain*. The gray matter between my head.